Water and Agriculture
in Colorado and the American West

Water and Agriculture in Colorado and the American West

First in Line for the Rio Grande

David Stiller

UNIVERSITY OF NEVADA PRESS *Reno & Las Vegas*

LIBRARY OF CONGRESS CATALOGING-IN-PUBLICATION DATA
Names: Stiller, David, 1947– author.
Title: Water and agriculture in Colorado and the American West : first in line for the
    Rio Grande / David Stiller.
Description: Reno ; Las Vegas : University of Nevada Press, [2021] |Includes
    bibliographical references and index. | Summary: "Water and Agriculture in
    Colorado and the American West: First in Line for the Rio Grande is a chronicle
    of drought and water shortages throughout the rapidly growing American West,
    where long-established agricultural water rights play an increasingly critical role
    in problematic attempts to satisfy agricultural and urban needs in the region"
    — Provided by publisher.
Identifiers: LCCN 2020048864 (print) | LCCN 2020048865 (ebook) |
    ISBN 9781948908801 (hardback) | ISBN 9781948908818 (ebook)
Subjects: LCSH: Water in agriculture—Colorado—History. | Water in agriculture—
    West (U.S.)—History. | Water in agriculture—Rio Grande Watershed (Colo.-
    Mexico and Tex.)—History. | Water-supply, Agricultural—Colorado—History. |
    Water-supply, Agricultural—West (U.S.)—History. | Water-supply, Agricultural—
    Rio Grande Watershed (Colo.-Mexico and Tex.)—History. | Rio Grande (Colo.-
    Mexico and Tex.)—Water rights.
Classification: LCC S494.5.W3 S748 2021 (print) | LCC S494.5.W3 (ebook) |
    DDC 631.709788—dc23
LC record available at https://lccn.loc.gov/2020048864
LC ebook record available at https://lccn.loc.gov/2020048865

First Printing
Manufactured in the United States of America

25 24 23 22 21   5 4 3 2 1

For Jean,
who makes the trip a grand adventure.

Contents

Illustrations

Maps and Illustrations (*following page 81*)

Map of the Rio Grande from Colorado to the Gulf of Mexico
Map of San Luis Valley and Upper Rio Grande
Map of Ten Major Irrigation Reservoirs Surrounding the
     San Luis Valley
Map of Primary Irrigation Canals, Laterals, and Ditches in
     San Luis Valley
San Luis Peoples Ditch
Community of San Luis and Culebra Creek Valley
Rio Grande Canal Diversion Dam on the Rio Grande
Rio Grande Canal at Headgate
San Luis Valley Irrigation District Office in Center, Colorado
Rio Grande at Southern End of San Luis Valley
Rio Grande Reservoir
Closed Basin Drain and San Luis Valley
Center Pivot Sprinkler, San Luis Valley
Alfalfa Stacks, San Luis Valley
Water for Sale, San Luis Valley

PREFACE

Not many years ago, my wife and I purchased a small farm in a western Colorado valley to pursue a new life growing nursery plants, alfalfa, and grass hay. The days were long and the work hard but, once accustomed to the hours, we settled in. One of our earliest challenges was adapting to the semiarid climate. Rain and snow on the valley floor were sparse, which was not quite desert but close enough that the difference seemed arbitrary. To deal with this, our farm, like all farms in the valley, required irrigation—the human transfer of water from where it is to where it is wanted in order to grow crops. Irrigation is essential for nearly all farming in Colorado and the dry American West.

The source of our irrigation water was a river that served as the valley's economic centerline. During spring runoff, when melting mountain snows fed the river, I listened to the river's noisy churn and noted the snow-draped peaks crowding our valley, and it was easy to dismiss local claims that the valley suffered from a water shortage. Then August came, and the riverbed near my farm was dry. Up-valley irrigation diversions were responsible; the sum of those diversions exceeded the flow of the river. I learned it was not unusual during late summer, when the natural flow of the river was declining and every irrigation diversion in the valley was wide open, for reaches of the river to go dry. There was nothing natural about it.

Over the next decade, if initially to myself, I began to question how the valley's water was used. Irrigation was neither off-putting nor unusual. Humans have irrigated for thousands of years. Nonetheless, and although more efficient forms of irrigation like center-pivot sprinklers were becoming common, I concluded that water was diverted from the river as much by habit as by immediate need. Water could flow willy-nilly across and alongside county roads, or occasionally reach the end of a canal or ditch, without being applied to a field, then drain unused into the nearest draw. I never heard an irrigator claim he had enough water, but it was easier and cheaper to squander water than to temporarily decrease a diversion. Water that was free a century ago was still free, or nearly so. Convenience and efficiency were interchangeable, defined in the moment. It was a short step for me to wonder how the valley's water habits came into being.

The result of my perplexity is this book, and my effort to answer how agriculture came to dominate the river in my valley and so many others in the West. It is no secret that the American West is running short of water, yet agriculture has long maintained the upper hand in its distribution and use. I have always been fascinated by the nexus between the human and physical environments, so pursuing the connection between water development and agriculture and how they co-evolved in the West became irresistible. Consequently, this is an account of water's rapid development and the subsequent competition, misuse and overuse, multistate bickering, and, eventually, tepid cooperation, all in pursuit of a finite and variable resource. It is a narrative of how transforming a major river to facilitate irrigation affected and changed the landscape, sometimes minimally and barely noticeable, occasionally in a major way. Finally, it is an overview of other Colorado rivers and how their stories are paralleled elsewhere in the West.

Curiosity may have prompted this book, but I understand how water moves through the environment. Before farming, my career had been in hydrology. I had decades of experience, strengthened by several graduate degrees in the earth sciences. I spent three years on the board of a corporation responsible for the delivery of irrigation water to ten thousand acres of farmland, which gave me the opportunity to observe valley farmers and how they addressed irrigation needs and infrastructure. And I directed a nonprofit organization whose mission was to improve river health, track water quality, and assist farmers with the construction of improved irrigation diversion structures.

Experience and curiosity aside, however, I questioned whether my home valley was representative of irrigation and water use elsewhere in Colorado and the West. Its watershed was small and its contribution to Colorado's agricultural output was a statistical rounding error. I had to find a river and valley better matched to my objectives. Eventually, I chose to undertake a case study of a much larger river valley where agriculture was the major economic force, where water and agriculture were intertwined without competition from urban or industrial needs. Colorado is a headwaters state. No major rivers enter the state; they only leave. Its mountains are the source of four of the nation's major waterways: the Arkansas, Platte, Colorado, and Rio Grande. Following weekends driving along and studying these four rivers, the Rio Grande and San Luis Valley drew my focus. Of Colorado's major rivers, the Rio Grande is likely the most heavily committed to irrigation within the state.

My research began in state and city libraries in Denver and the University of Colorado. I completed hours of interviews of local and state water officials, scientists and engineers, attorneys, and knowledgeable stakeholders and water users in the San Luis Valley and Denver on topics that included irrigation technology, water law, and valley history. Getting to know these individuals was the most enjoyable and interesting part of the process. Today, I consider many of them friends. And I travelled incessantly through the San Luis Valley, observing irrigated fields and cropping systems, diversions, canals and ditches, and reservoirs located both in the valley and in the surrounding mountains. I do not regret a single day spent researching and writing this book.

This story covers the broad historical arc of the upper Rio Grande and Colorado's San Luis Valley, from initial exploration in the first quarter of the nineteenth century, to the arrival of Spanish farmers in the 1850s, followed by rapid Euro-American settlement in conjunction with Colorado's 1858 gold strike. The valley has always been dominated by agriculture. Almost from the beginning, agriculture in Colorado was influenced by the legal Doctrine of Prior Appropriation as the basis for its water law. Withdrawing irrigation water from the Rio Grande accelerated farming in the valley, although not without periodic and serious mistakes in water use, high-handed treatment by Washington, and the need for construction of numerous irrigation reservoirs early in the twentieth century. International and interstate squabbles over how to share the river eventually led to creation of the Rio Grande Compact, although Colorado failed to abide by it until the U.S. Supreme Court's involvement was threatened. Improvements in irrigation technology and discovery of an untapped source of water allowed a near-continuous increase in both the number of acres under irrigation and the amount of water used. And recent proposals to export valley water to Colorado's urbanizing Front Range cities forced valley farmers to organize and cooperate rather than bicker. Through all of this, periodic drought created havoc and anxiety in the valley.

I hope this book fosters a better and broader understanding of the origin and impact of the agricultural practices that determine how the

Rio Grande's—indeed, the West's—water is used. Agriculture consumes eighty to more than ninety percent of the West's water. In the San Luis Valley, it may be more than 95 percent. Stated differently, the West's cities and all other uses account for less than twenty percent of the region's total water consumption. The West's population is growing, while its water supply is not. Notwithstanding constraints imposed by established water law, it will be up to the West's farmers and non-farmers to settle on how much irrigation remains a worthwhile use of a limited public resource. If current agricultural and irrigation practices are unsustainable, nothing will save western irrigation from change.

And finally, I trust that this account fairly describes the farmers and ranchers of the San Luis Valley, Colorado, and the West, and how they have developed and continue to use the region's finite water resources. I have endeavored to fairly mirror their practices and communities. I especially hope this informs those inhabitants of Colorado and the West who do not farm, do not raise livestock, and do not irrigate, yet are becoming increasingly concerned that the West's water is limited.

Water and Agriculture
in Colorado and the American West

CHAPTER ONE

The River, the Valley, and Early Customs with Water

The basic purpose of Mexican law was not to stimulate private enterprise but to irrigate the maximum acreage... [W]herever possible irrigation should be a community endeavor...

— Donald J. Pisani, *Enterprise and Equity: A Critique of Western Water Law in the Nineteenth Century* (1987)

At the south end of Colorado's Rocky Mountains, just west of the small community of Del Norte, the *Rio Grande*—Spanish for Great River—hurries past a final rock buttress and pours into the San Luis Valley. In the process, it changes from mountain cataract to meandering river. Above Del Norte, the Rio Grande is confined to a canyon carved into the San Juan Mountains. Below Del Norte, the river meanders across a valley fifty miles wide and almost twice as long, an earthly depression so broad and distinct as to be noticeable from space. After leaving the San Luis Valley and Colorado, the Rio Grande passes for hundreds more miles through New Mexico and around Texas to the Gulf of Mexico. Along its path, it waters hundreds of thousands of acres of farmland.

Some of Colorado's tallest and snowiest mountains feed the Rio Grande and the San Luis Valley, but at nearly eight thousand feet above sea level the valley floor is desert. Meaningful rain is sparse; total precipitation averages less than eight inches a year. Snow, when it falls, is dry as powder. Sunshine is the norm, yet the air is cool and dry on the skin. Winters are harsh when cold air from the surrounding mountains slides into the valley. Longtime valley residents speak of "30–30" winters: thirty inches of snow and minus-thirty degree temperatures. When mountain snows finally melt, from April through June, the Rio Grande surges to a swift current of foam, detritus, and waves. Three-fourths of the Rio Grande's annual flow occurs during those three months. In late summer, as mountain snows disappear, the river's flow decreases through the fall and winter until mountain snowmelt begins once more the following spring. It is an annual cycle replicated by rivers throughout the West.

The Rio Grande has been entangled in the geology of the San Luis Valley for hundreds of thousands of years. During this span, the river intermittently spilled into the valley in rapids and riffles or emptied quietly into prehistoric lakes that once covered much of the valley floor. Now, the river is free to meander between where it enters and exits the valley, a distance of roughly seventy-five miles. As grand as the river has been and is, however, it is undersized for the valley. Thirty to thirty-five million years ago, huge blocks of the earth's crust underlying what is now the American Southwest began to lurch, twist, and sag. Part of the resulting deformation became the Rio Grande Cleft, a colossal sub-surface trough that extends from the north end of the San Luis Valley to near El Paso, Texas. As the floor of the San Luis Valley slowly sagged between five and nineteen thousand feet, it was backfilled with sediment shed from surrounding mountains.[1] The valley came first, then the river. The accumulated layers of gravel, sand, clay, lava, and volcanic ash created what one can envision as an immense bathtub full of sand and gravel and then saturated with water. The resulting *aquifer* contains billions of acre-feet of stored groundwater.[2]

Step forward to two hundred years ago and the Rio Grande's path could be seen from miles away as a snaking corridor of cottonwood and boxelder trees that lined the river's banks and covered its floodplain. Willow thickets hugged the river, providing cover for waterfowl and shade for fish. Thousands of ducks, geese, and sandhill cranes called the San Luis Valley their permanent home, and tens of thousands used it for stopovers on their twice-yearly migrations. Between the surrounding mountain slopes and the river were miles of dryland sage and rabbitbrush, lush meadows, and extensive wetlands. Nothing in the valley at the time hinted at permanent civilization. Native American Utes moved through the valley with the seasons.

Major Zebulon Pike's name is little known outside Colorado. Within the state, however, his moniker is the well-known precursor to Pike's Peak, the prominent fourteen-thousand-foot granite massif bulging from the Front Range, the first row of mountains visible from the Great Plains. In 1806, just months before Lewis and Clark returned from their famous wilderness epic, Pike set off on the nation's second official exploration of the West. He travelled westward up the Arkansas River and into the mountains, eventually entering the San Luis Valley, in the process unknowingly leaving the recently acquired Louisiana Territory and entering Spanish territory. In early 1807, after being intercepted by a Spanish army detachment along a tributary to the Rio Grande, he was escorted to Santa Fe and interrogated. Pike and his men subsequently were allowed to return to the United States.

The United States undertook its next formal exploration of the West in 1820, placing the effort under the command of Major Stephen Long. Long travelled up the Platte and South Platte, south along the Front Range, and then returned east down the Arkansas River. He did not enter the San Luis Valley but, similar to Pike, one of the interesting and important outcomes from his expedition was his map. Where Pike had established the longitude of the Southern Rockies, Long accurately fixed the latitude of notable features. Cartographers finally were able to accurately locate the Southern Rockies on their maps, an indication that Americans were uncertain of the precise location of the Rocky Mountains until after 1820. Prior to then, their location was essentially so many days' travel westward across the Great Plains.

Pike's and Long's expeditions coincided with the finest years of the Rocky Mountain fur trade. Trappers and traders began wandering through the young American West before 1810, but their interests were more commercial than exploratory and their few surviving records are not terribly informative.[3] After Mexico gained its independence from Spain in 1821, the San Luis Valley became part of Mexico, although ownership was of little concern to the trappers and traders who used the valley as their thoroughfare to and from Santa Fe. As the fur trade faded in the 1830s, there were no American habitations in the valley or along the upper Rio Grande. But major changes were coming.

◆ ◆ ◆

In the San Luis Valley's southeast corner, surrounded by irrigated hay-fields and back-dropped by the Sangre de Cristo Mountains, lies the tidy community of San Luis. It is Costilla County's largest town and the county seat. In the customary grid of western settlements, its paved and dirt streets point in the cardinal directions. Cottonwoods overhang stucco and wood-frame buildings. The largest structure, a stone-and-mortar Catholic church, centers the town. Across the street sits R & R Grocery. Opened in 1857, it is Colorado's oldest continuously operated store. San Luis residents are quick to emphasize their Spanish, not

Mexican, ancestry. Seven-eighths of its residents are Spanish American, and Spanish surnames label numerous streets. Many inhabitants trace their families' presence in the area to the time before Colorado statehood.

At the town's south end a small stone monument stands by the highway. Water runs in a ditch that passes under the road in a culvert. Irrigation ditches are as common as rocks in Colorado, and differentiating one from another can be difficult for an outsider. Perhaps six feet wide and three feet deep, this one would go unnoticed but for the monument and a weathered bronze plaque that identifies the ditch as the San Luis Peoples Ditch, bearer of Colorado's oldest water right: Court Decree Priority No. 1, dated April 10, 1852. Across the highway a short, lean man wearing a faded blue denim shirt operates a backhoe. He stops the machine, walks over, and confirms that the ditch is indeed the San Luis Peoples Ditch. His accented English contrasts unexpectedly with piercing blue eyes.

It is October. Outside San Luis, hayfields and pastures form an irregular lattice within a network of cottonwood-lined streams and more irrigation ditches. The hayfields have been cut for the last time this year, and the stubble is brown and crisp. Single-story frame and stucco homes, pole corrals, and gardens occur in scattered clusters. A mile above San Luis, a small dam diverts twenty-one cubic feet per second (CFS) of water from Culebra Creek into the Peoples Ditch. Twenty-one CFS may seem a trivial flow, but it is a substantial quantity: 9,450 gallons of water every minute. The greater San Luis Valley has hundreds of irrigation channels like the Peoples Ditch. Many are larger, some are smaller, none are older.

The valley's first permanent settlers, Spanish *pobladores*, migrated up the Rio Grande from near Taos, New Mexico Territory, in 1849. They made their first effort at a settlement about twenty-five miles southwest of San Luis. More families arrived two years later and settled along Culebra Creek. Utes chased them back south before winter, but they returned for good in 1852. All were part of fifty families selected to settle the nearly one-million-acre Sangre de Cristo Land Grant, controlled by Carlos Beaubién, a Canadian-born, naturalized Mexican citizen. After Mexico lost the Mexican-American War in 1846, Mexico ceded what would become the American Southwest, including the San Luis Valley, to the United States in 1848. Beaubién managed to affirm his land grant under American law; thus, whether aware of it or not, the pobladores were settling in the United States. Within days of their arrival on Culebra Creek, they began to construct a community that would become San Luis. The settlement was one of several that sprang up along Culebra Creek in the 1850s, as well as the first organized town in what would become Colorado Territory in 1861. San Luis and its neighboring communities have remained a Spanish-speaking center in the San Luis Valley. Spanish is, if not the first language a newborn hears, the second.

The same year that they constructed their fortified community, or plaza, the settlers also began to excavate the Peoples Ditch, their *acequia*, to divert water from Culebra Creek to their fields. Acequias were more than irrigation ditches. They were also the embodiment of a self-governing, communal tradition that provided water to multiple family farms and gardens through collective construction and maintenance. As with nearly everything else in their lives, their construction methods were brought with them from Taos. They would have first excavated a hole in the bank of Culebra Creek, then scraped and dug, using wooden shovels and hand implements, a ditch angling downvalley and away from the creek. When wooden shovels and hoes were inadequate, or they wanted to enlarge the ditch, the pobladores would have hitched oxen to wooden plows. As the ditch was angled away from the stream, the amount of land available for irrigation expanded as the distance between the stream and the ditch increased. Prior to the Peoples Ditch, Culebra Creek naturally wetted only a narrow riparian zone along the stream that rarely exceeded a hundred yards in width. The purpose of the acequia was to spread water over a broader area, and irrigated fields eventually extended nearly two miles across the Culebra Creek Valley. The creek was fed largely by melting snow in the Sangre de Cristo Mountains, but available runoff only went so far, and irrigating fields in the upper watershed was accomplished at the expense of fields downstream. During the irrigation season, there was almost never water in Culebra Creek for the final ten miles before it joined the Rio Grande.

Once diverted, water would have flowed by gravity to the settlers' individual parcels, where it was redirected to their fields, gardens, and orchards. The pobladores immediately planted beans, chilies, white corn, wheat, oats, onions, field peas, and pumpkins, as well as apple and plum saplings brought from Taos. Farming along Culebra Creek was further encouraged by a flour mill constructed near San Luis. By

1860, the communities along Culebra Creek were home to nearly eighteen hundred residents, nearly all Spanish farmers and herders from the Taos area.[4]

Following the settlement of San Luis and construction of the Peoples Ditch, the next half-dozen acequias quickly followed, all in the Culebra or nearby Costilla Creek drainages.[5] By 1882, at least twenty-three acequias were operating just within the Culebra Creek watershed.[6] Across the Rio Grande to the west, other Hispano settlements sprang up on the neighboring Conejos Land Grant. The town of Guadalupe was settled in 1854, followed by others farther north along La Jara Creek and the Alamosa River, and more acequias were excavated.[7]

It was no accident that the valley's first settlers chose perennial mountain streams on the valley's perimeter for their farms. The first to settle, they claimed the best land. The Rio Grande was too boisterous for their tools and skills. Land next to the river would have been avoided such that they would not have to reclaim their fields and reconstruct their irrigation infrastructure after every spring flood. Placing their farms higher on the valley sides avoided this likelihood. They also benefitted from slightly longer growing seasons. Winter's cold mountain air settled in the valley's lowest areas—along the Rio Grande. Spanish was the spoken language, not English. Farming and survival were what mattered, not language or citizenship.

By the mid-nineteenth century, the Taos Valley in New Mexico Territory had been completely broken into small plots and farms. Consequently, as soon as Spanish farmers learned of the farming possibilities in the San Luis Valley, there was solid reason to leave the Taos environs and resettle in Colorado's higher climate. To protect Spanish settlers from the Utes, the U.S. Army constructed and manned Fort Massachusetts in 1852, then a better-located Fort Garland farther south six years later. After the Utes were finally forced out of the San Luis Valley in the late 1860s, pobladores began settling in decentralized plazas rather than more defensible towns, preferring to be closer to their fields, water sources, and acequias.[8] A dozen families built a new plaza near Del Norte, nearly sixty miles northwest of San Luis. Almost as quickly, plazas and settlements extended up the valley's west side across a landscape of mixed grasslands and forested slopes interspersed with saltbush, shadscale, sagebrush, and greasewood shrublands. Like those who settled the valley's south end, these later settlers almost uniformly farmed along tributaries exiting the surrounding mountains. "Spanish-speaking settlers had by the early 1860s transformed the valley into an area typical of rural northern New Mexico, with small plazas dotting most of the land which could be irrigated easily."[9]

The Spanish settlers who moved into the San Luis Valley may have broken new ground and created new farms, but their irrigation practices were anything but original. When the skies fail to provide reliable and adequate growing season precipitation for food crops, cultures and civilizations develop the practices to move water to where it is needed. They applied the same irrigation practices originally brought to Spain by North African Moors. The word *acequia* has an Arabic root, not Spanish or Latin. Moorish practices were subsequently adapted over many years before they were imported to the Western Hemisphere by Spanish colonists. It was only a matter of time before these practices found their way to Taos and then the San Luis Valley.

The fundamental characteristic of all acequias was the sharing of available water. From the excavation of their acequias to the way they were managed, everything was based on the communal nature of their settlements.[10] Sylvia Rodríguez, an ethnographer who has studied acequias in northern New Mexico's Taos Valley, has written, "By definition, water is always shared, sometimes simply and sometimes in more complicated ways. The tacit, underlying premise is that all living creatures have a right to water."[11] In times of water shortages—typically beginning after spring runoff and lasting into the fall—Spanish colonists shared according to need and equity. Water and the right to use it was not something owned exclusively by individuals. The principal of sharing ruled. Over time, the division of water in a river or acequia became a custom, or *reparto*. Each poblador who was awarded and settled a tract of land and assisted in constructing the acequia became a *parciante*, in effect a sort of shareholder on the ditch and thus entitled, even required, to participate in its operation and management. Under the acequia system, water and land went together and were not transferred or sold separately.[12]

Each spring, a *mayordomo* was appointed to manage the acequia's operation and division of water for the year. Perhaps the most important function of the mayordomo was to implement the agreement among parciantes on how to divide available water during times of drought and water shortage. His decisions held, the most common of which authorized him to manage individual headgates on the acequia so that water was delivered on the basis of immediate need, especially for livestock, gardens, and orchards. A parciante could receive a steady amount of water for a fixed period of time, a comparative trickle of water for a longer period, or some variant of the two. Once a parciante's water passed through his headgate to his land, the actual application of water to crops began. Applying water to dry soil in pastures, gardens, and orchards would seem to be an undemanding task, but as with most aspects of nineteenth-century irrigation, it was both arduous and skill-based. Farmers used primitive hand tools to flood a field by maneuvering water over ground that was never perfectly level.

Mayordomos were similarly tasked with reaching parallel agreements with their counterparts on other acequias, drawing water from the same river or stream at the same time. Neighboring mayordomos negotiated how to share entire rivers. The practice of negotiating water shortages among acequias diverting from a common stream continued until Colorado implemented a very different way of apportioning water in the 1860s, although the Colorado Territorial Legislature enacted a law in 1866 authorizing farmers in Costilla and Conejos Counties—at the south end of the San Luis Valley—to continue their acequia and mayordomo traditions.[13]

Today, acequias remain a community-based means for distributing irrigation water to ranchers and farmers in Costilla and Conejos Counties. Although their customs have been modified to adhere to twenty-first-century Colorado water law, and the tools used in those first decades have been replaced by steel implements and diesel power, the original principles of acequia culture still apply. In Costilla County alone, more than one hundred acequias continue in use.[14]

The impacts of nineteenth-century Spanish settlements on the Rio Grande and San Luis Valley's environment were relatively minor. Collectively, however, these small settlements characterize how the earliest agricultural communities used water in Colorado. Within the next several decades, a wave of Euro-American settlers, motivated by different customs and beliefs, would arrive. Many, perhaps most, would come from more humid parts of the United States and would be ignorant of irrigation. Nonetheless, they would adapt and significantly change how and where water was used, in the process participating in the creation of a new ethos toward water and water law that would alter the San Luis Valley and Colorado in dramatic ways.

CHAPTER TWO

MINING AND FARMING—ONE FED THE OTHER

*No industry had a greater impact on Western history
than did mining.*

— Patricia Nelson Limerick, *Legacy of Conquest* (1987)

Boosters often wrote as much about farming as mining.

— Elliott West, *The Contested Plains: Indians,
Goldseekers, and the Rush to Colorado* (1998)

In 1838, as trappers watched their era come to an end and began to
settle onto farms in pockets of the West, the U.S. Army undertook
a new mission that would have far-reaching consequences when
Congress authorized formation of the Corps of Topographical Engineers. What made this corps unique was its role in mapping the American West. Before its creation, limited mapping had been performed by
the better-known Corps of Engineers. Congress recognized, however,
that with new land to explore, settle, and exploit, Americans would
begin a wave of westward movement for which accurate maps were
indispensable.

The new maps compiled by the Army were to scale; every physical
feature transposed onto a map was correctly placed relative to every
other feature. The cardinal directions were precisely labeled; map scales
and distances were accurate. The latitude and longitude of landforms,
river bends and stream courses, heights of mountains, and the location
of towns and mining districts were all plotted so that the Army, or anyone with a compass and a little sense, might find his or her way from
point A to point B, or possibly from the Missouri River to the Rocky
Mountains or Pacific Ocean.

Captain John Gunnison was a member of this new corps. His name ultimately would be repeated throughout the Rocky Mountain West: streets, towns, and rivers bear his name in the twenty-first century. In 1853, he was assigned to reconnoiter a potential railroad route across the Southern Rockies. He passed through the San Luis Valley that summer, but where Zebulon Pike's reflections on the valley were shaped by winter snow and cold in 1807, Gunnison's observations were detailed and illuminating. Although he was later killed by Paiutes in Utah, his second-in-command, Lieutenant Edward Beckwith, finished the survey and wrote the final report based largely on his notes.[1]

On August 13, 1853, Gunnison's party of loaded wagons and horsemen descended into the San Luis Valley, camping just below Fort Massachusetts, which had been constructed one year before. While most of the party rested, Gunnison and several others spent several days exploring the rugged country to the south. His travels proved to be so arduous that he grasped the value of a knowledgeable guide and ordered Beckwith to Taos to hire one. Beckwith's path to Taos followed Sangre de Cristo Creek into the San Luis Valley, then turned south, where the valley was "traversed by the Rio Grande del Norte and its mountain tributaries, skirted with bushes and a little timber."[2] They crossed Culebra Creek five miles below a new Spanish settlement named San Luis and continued on to another new community, Costilla, where "a few fields are already covered with crops of corn, wheat, oats, and the other usual crops of a New Mexican farm."[3] During his ride of over one hundred miles from Sangre de Cristo Creek to Taos, Beckwith "saw no grass in the valleys worth naming; the vegetation being confined almost exclusively to *artemisia* [sagebrush] and a few varieties of cacti, but chiefly the prickly pear." He opined that crops could not be grown in the valley without irrigation.[4]

After hiring a guide, Beckwith rejoined Gunnison's party as it headed north along the San Luis Valley's eastern edge. "The grass along our path was scattered, and we experienced considerable difficulty in driving over the thick masses of sage which cover almost the entire surface of this immense valley."[5] As they approached what today is Great Sand Dunes National Park, their new guide told them what he knew about the valley:

> to the west of our trail, along the banks of the Rio del Norte to where it enters a plain through a cañon from the San Juan mountains…the valley of San Luis…is rich and fertile, covered

with extensive meadows of grass.... The narrow line of timber, thirty-five miles distant upon the Rio del Norte, is plainly seen from our trail; but it is represented to be difficult to cross the valley with wagons, on account of the marshes along the river and the miry banks of the sunken creeks.... The San Luis valley is...so level that trees are seen in any direction, growing on streams, as far as the eye can discern them.[6]

Continuing north, Gunnison's party fought through more sagebrush and was forced to dig in streambeds for drinking water, because streams emerging from the mountains surrounding the valley promptly disappeared into the valley floor. "To our left we could see fine prairie-grass fields, directly in the course to the Coochetopa Pass...but the guide warned us of marshes and the attempt was not made to cross them.... The surface of the ground, over large spaces, is often covered with effloresced salts."[7] Common in the arid West, these natural salts would prove significant to a developing agricultural economy over the next century.

Gunnison left his party so he and several others could travel up San Luis Creek, where he discovered "the most luxuriant fields of grass seen on the trip," and where "much hay could be cut, and fine grazing farms opened; and it is also probable that wheat and flax, and perhaps other grains, could be raised."[8] After Gunnison rejoined Beckwith, they crossed Coochetopa Pass and continued westward on their mission.

Beckwith's final report delivers a strong opinion about the valley's attributes and deficits. "We here leave the immense valley of San Luis, which is one of the finest in New Mexico [sic], although it contains so large a proportion of worthless land—worthless because destitute of water to such an extent where irrigation alone can produce a crop, and because of the ingredients of the soil in those parts where salts effloresce upon the surface."[9] Along their route, true to their guide's admonitions, a discontinuous, shallow water table and poor drainage created marshes and wetlands they had to avoid by backtracking. They discovered a "worthless land" that might be suitable for agriculture, but only after farmers dealt with saline soils and the shortage of water. Their description of the Rio Grande was only as a tree-line observed from miles away. And *irrigation* became a new word in the American lexicon.

◆ ◆ ◆

In 1858, change in the San Luis Valley accelerated to a magnitude unforeseen by Beckwith just five years before. It began several hundred miles to the north with a gold strike on the South Platte River. More immigrants subsequently headed for the new gold fields in 1859 than had journeyed to California ten years before.[10] The new Pike's Peak or Bust phenomenon encouraged as many as one hundred thousand would-be gold-seekers to leave their farms, jobs, and families back east and cross the Great Plains to try their hand at goldmining.

The prospectors and miners who passed through the San Luis Valley in the 1860s primarily accessed the rugged high country in search of precious metals by following rivers and streams, one such drainage being the Alamosa River, which drains the San Juan Mountains into the San Luis Valley roughly midway between where the Rio Grande enters the valley and where it enters modern New Mexico. The Alamosa was not a river, but a stream with a river's name; little of its flow ever reaches the Rio Grande today. From the valley floor, the Alamosa climbs westward more than thirty miles to its headwaters of ice-water seeps and springs along the Continental Divide. Together with nearby La Jara Creek, the Alamosa was one of the first streams along which Spanish settled on the valley's west side in the 1850s and early 1860s.

In late June of 1870, five prospectors working their way up the Alamosa River into the mountains came to a tributary that showed signs of placer gold. They followed the tributary, eventually naming it Wightman Fork after one of the prospectors, panning and sluicing their way up the canyon bottom. The Alamosa River and its tributaries flowed in "deeply entrenched valleys above which the mountain slopes rise with unusual steepness to elevations above the valley bottoms of one thousand to twenty-five hundred feet."[11] Most slopes were covered with thick forests alternating with strips of meadow and stands of jack-strawed trees downed by fire and wind. Beginning in the early 1860s, large portions of the region's forests burned. Early prospectors and miners were cavalier about controlling their campfires, a careless habit repeated throughout the West during mining's nineteenth-century heyday. Charred and wind-mangled forests made access difficult, yet the attraction was there: "the brilliant red, yellow and brown coloring of the mountain slopes" created by the physical and chemical weathering of wealth-bearing rocks and minerals.[12]

Amidst vivid scenery, the prospectors spent the summer of 1870 exploring and mining in upper Wightman Fork. Eventually, they approached the grassy swale and slopes of South Mountain, where they quickly identified gold- and silver-bearing lode deposits. Snow forced them back to the San Luis Valley for winter, but the following spring they returned, accompanied by abundant friends and aspiring gold-seekers. Numerous lode claims were staked during the new mining district's first years, including those that became the most successful mines on South Mountain, the mountain's name providing evidence that prospectors compensated for their industriousness with a lack of imagination. By 1873, the area was in year-round operation. Access to the new mining camp, called Summit, was through the Alamosa River canyon, a difficult route. An easier toll road was eventually constructed twenty-seven miles over the mountains to a new town, Del Norte, located on the Rio Grande.

By 1876, the town of Summit boasted about five hundred residents, plus stores, saloons, and a post office. In 1880, it was renamed Summitville. Many of the district's mines were ultimately connected by a three-dimensional maze of adits, winzes, raises, crosscuts, and tunnels, all used to access ore, haul it and waste rock from the mountain's interior, and drain the mines of groundwater. Stamp mills were hauled piecemeal to Summit by late 1874; nine of them clanged noisily in the district by 1883. Four tramways transported ore from the mines to the mills.[13] The mill waste, called *tailing*, was flushed down Wightman Fork to the Alamosa River.

Experience in the Summitville Mining District, along with gold fever, one of humankind's most contagious diseases, enhanced the region's learning curve as prospectors continued to scour the complex volcanic rocks comprising the San Juan Mountains. Another gold strike occurred lower on the Alamosa River in 1874, from which the town of Jasper arose. More strikes occurred in the vicinity and more towns popped up, including Stunner, Blainville, Loyton, and Platoro. All boomed intermittently from the late 1870s into the 1890s. Ultimately, only the Summitville Mining District produced significant amounts of high-grade ore.[14]

The frenzy ended almost as abruptly as it began. By 1895 most Summitville mines had closed and activities were limited to the few optimistic and stubborn individuals willing to scratch away for very little or nothing. During those first twenty-five years, the order of one million dollars in gold and silver was extracted—a vast fortune in nineteenth-century terms. The figure is an estimate; no one kept records. No records, no reports were required for minerals extracted from federal land, and no royalties were paid to the federal government for the wealth extracted from the public estate.

Summitville and the upper Alamosa River watershed were not the only areas near the San Luis Valley that experienced gold and silver strikes. More than thirty miles northwest of Summitville, a major silver strike occurred near Creede, on Willow Creek, a tributary to the Rio Grande mainstem, in 1869. The ore proved difficult to process, however, and mining did not begin to flourish there until technological improvements to ore processing prompted the onset of serious mining around 1889. North of the San Luis Valley, another silver strike prompted mining at Bonanza in 1880. Other mineralized areas adjacent to the valley enjoyed less notoriety, including Crestone in 1879 and a gold strike above San Luis in 1890. These latter two strikes occurred in the Sangre de Cristo Range, east of the San Luis Valley. The area's silver mining centers initially boomed, prospered, yet faded rapidly following the Silver Panic of 1893. Unaffected by the silver panic, goldmining in the mountains surrounding the San Luis Valley continued into the twentieth century.

The search for gold drew immigrants to Colorado like picnics draw ants. Prospectors and miners whirled through the South Platte's headwaters and into the surrounding mountains and valleys, including the San Luis Valley. Yet hundreds gave up and returned home. "Go-backs," they were called. Of those who stayed, many decided that they either did not know enough about prospecting and mining to continue, or that they could make a better living from the soil. The farming and animal husbandry skills that many brought with them could be put to good use. The mining history of the Southern Rockies has been well documented, but what has been too-often overlooked or underreported is that every prospector, miner, millworker, and anyone else living and working in the mining camps was accustomed to eating on at least a semiregular basis. Their food had to come from somewhere. Pony Express deliveries of hot meals from St. Louis or Kansas City were not an option.

The need to feed miners and millworkers in the mining camps helps explain the origin and growth of farming and ranching in Colorado and the West. Just as the South Platte Valley became home to farms and ranches to feed the mining camps in the mountains above Denver, so too did the San Luis Valley feed the mining camps in the San Juan Mountains and Sangre de Cristo Range. As the mining industry dug up streambeds, sunk shafts, and built mills and smelters, farmers began staking out farms, initially along streams and rivers on the plains east of the Front Range, but also in the San Luis Valley. Spanish settlers around San Luis may have been the first to farm in the San Luis Valley, but within a dozen years, spurred by the federal Homestead Act of 1862, many Euro-Americans began to join them.

By the end of 1859, the mining era's first full growing season, farmers throughout what would become Colorado Territory were providing produce and meat to the mining camps. The camps were never satisfied with the availability or the prices they paid for food, while the farmers and merchants were constrained by weather, available water, and the long distances over which they had to transport their products. Both groups were dependent upon each other, yet neither was fully satisfied.[15] Mining both triggered and stimulated the development of the San Luis Valley's agricultural economy, as well as that of Colorado. The state's birth is tied to agriculture as much as it is to mining. And things were just beginning.

CHAPTER THREE

A DOCTRINE FOR WATER TAKES SHAPE

As settlement of the territory intensified, agriculture began to replace mining as the sector principally concerned with and responsible for the use of water.

— David Schorr, *The Colorado Doctrine* (2012)

The discovery of gold in 1858 occurred not in Colorado Territory, which did not exist yet, but in Kansas Territory. And if one was prospecting in the headwaters of the Rio Grande, he did so in New Mexico Territory. In fact, a fistful of territories encompassed the American West at the time, including Utah, Nebraska, Oregon, and Washington Territories. California was admitted to the union in 1850, but nearly everything else in the West had yet to arouse much Congressional attention. The national debate and unruliness that would lead to the Civil War in 1861 were already consuming the political oxygen. Deciding how to govern and provide a federal presence in the West simply did not happen.

The prospectors and miners swarming through the mountains above Denver did not wait on Congress to provide civil guidance. Almost as soon as they arrived in the mud-and-timber mining camps, they began to negotiate and settle on systems of self-governance. Many of those who descended upon these camps either had experience in California or came directly from its gold camps, so it is not surprising that they transplanted many of California's habits and customs to what quickly became mining districts. Just as quickly, the districts became functioning democracies. They elected governing and administrative officers, established civic procedures and codes of behavior, and settled on how mining claims were to be staked and registered. Procedures on documenting legal ownership and transferring mining claims were particularly important.

Of comparable importance was how water was used and controlled. Most prospectors employed *placer* techniques, at least initially, where running water was used to separate the heavier native gold from sand and gravel in streambeds and canyon floors. Placer methods varied but almost all required plenty of water. If a claim was located on a flowing stream, the owner or operator was in an advantageous position. Problems arose when claims in need of running water were halfway up a dry mountainside. At first prospectors carried their diggings to water, but this approach was quickly discarded in favor of moving water to their diggings. Hundreds of ditches were excavated in and surrounding the mining camps and districts as flowing streams were diverted to mountainsides and other dry locations. This practice worked for many, but as mining districts became checkerboards of claims and diggings owned or controlled by multiple individuals, constructing a ditch across another man's claim generated unwelcome conflict and occasional bloodshed. District residents soon realized that a peaceful and orderly solution was necessary if they were going to spend more time mining than arguing and fighting. One useful approach was adopted in Gilpin County, in the mountains above Denver, in its 1859 Mining District Code: "It shall be the privilege of enny miner or Miners to take the water out of North Clear Creek in a ditch or floom around enny mans Claim or over his Claim for the purpose of washing Dirt on the Hill Side by Hydraulic power or slusing not ingering the claims passing thare over."[1]

Creative spelling notwithstanding, the protocol recognized the right of a miner to divert water from a stream across any intervening property to assist his operation. It copied practices adopted barely a decade earlier in California. And as with mining claims, the districts became the vehicle by which miners recorded the location and facts surrounding their ditches and water use.[2]

The right to divert water across another person's property was only half the battle. Of perhaps greater importance was the right to use or control water in the first place. Over the first two decades following gold's discovery, as the mining industry grew and Colorado came into existence, the process by which water would be apportioned and controlled evolved in important ways. In some districts, competing claims for the right to divert and use water would be resolved through sharing.

Gradually, however, a uniform practice was adopted whereby competing water claims were resolved in favor of priority, which simply meant that water was allocated in the order in which a water right had been claimed.[3] The emerging principle became known as "First in time, first in right." This phrase is still heard and used today.

Similar practices were adopted throughout the mountainous mining districts, all owing their origins to principles developed a decade earlier in the goldfields of California.[4] Collectively, they became the first legal statement of a new strain of water law known as the Doctrine of Prior Appropriation. In its strictest form, it would become labelled the Colorado Doctrine. Importantly, it replaced the much older, more widely used Riparian Doctrine, which dates to the Roman Empire and found its way to the United States through English Common Law. The Riparian Doctrine maintains that water in a stream belongs to the public for fishing and navigation and cannot be controlled by individuals. A property owner adjacent to a stream (a riparian landowner) has the right to make reasonable use of water in the stream, but may not obstruct the stream's use by others. If he diverts water from a stream, e.g., for a mill, he must return the water to the stream unchanged in quantity and quality. Significantly, all water users share a stream in times of drought.[5]

A moral dictate to share a limited resource like water certainly shaped the acequia culture of Spanish settlers in the San Luis Valley, and through their first formative years Spanish farmers changed their attitudes very little. Their irrigation practices originated with traditions brought from the Taos Valley, preceding those routines developing in the mining camps.[6] The San Luis Peoples Ditch and facsimiles in the San Luis Valley would eventually be caught up in Colorado's new water doctrine, but the two methods initially continued in parallel. When water was inadequate, the mining districts settled conflicting claims among miners and prospectors by applying the new doctrine that favored priority.

Colorado Territory's first Euro-American farmers had to adapt quickly to the reality that water was in short supply. Most came from the Midwest and East, where precipitation and streamflow were frequent and adequate and riparian water law prevailed. Their first observations probably led them to believe that western mountains were full of snow every winter and that streams gushed into the valleys and onto the plains every spring—until spring runoff ended and dry, hot summer days and weeks followed and the need for irrigation became apparent.

Irrigation was new to them. And as the mining camps grew and multiplied and small towns began to flourish, more farmers arrived, and more water was needed. Water shortages, exacerbated by nature's finicky ways, complicated farming in unexpected ways.

Eventually, agriculture began to take an increasingly important part in the territory's economy. Farmers on the plains around Denver chose not to adopt the sharing tradition employed with acequias, instead preferring to adopt the Doctrine of Prior Appropriation, initially along the Cache la Poudre and South Platte Rivers. By 1860, two short years after Colorado's first gold strike, thirty-five thousand acres were under irrigation along those two rivers and in the San Luis Valley.[7]

In 1861, Congress pieced together Colorado Territory from portions of Kansas, Nebraska, Utah, and New Mexico Territories.[8] Later that same year, Colorado's first Territorial Legislature, in an effort to remedy growing problems with agriculture and water, passed "An Act to Protect and Regulate the Irrigation of Lands." It stipulated that landowners located near a stream or river were entitled to use its water for irrigation and codified the right of any person diverting water for agriculture to a right-of-way for a conveyance ditch across intervening property regardless of the latter's ownership. This edict clearly upended the intent of riparian water law, which limited water rights to landowners adjacent to streams and rivers. The nascent Doctrine of Prior Appropriation, still in wobbly form, and derived as it was from necessity and based on at least the seasonal scarcity of water in Colorado, thus elicited the first legislative act related to water in Colorado Territory. It specifically benefitted irrigated agriculture and placed the territory's imprimatur on irrigation, a practice new to most settlers.

One of the most interesting aspects of the 1861 territorial law was how it mandated the division of water when there was insufficient flow in a stream or river to meet the wants of all users. In another surprise to advocates of riparian law, the new law mandated that when water claimants were in conflict over how much each was entitled to, the nearest justice of the peace would appoint three individuals to divide the streamflow in a "just and equitable proportion." These commissioners, as they were called, enjoyed substantial flexibility in how to satisfy this responsibility.[9]

Colorado's novel approach turned traditional riparian law on its head. But consider the environment and circumstances where it developed. Colorado was hundreds of miles from anything approaching the civilization whence nearly everyone had come. Natural resources were free for the taking, including water, even though the latter's abundance varied dramatically and seasonally. Everybody took what they wanted, limited only by the rules and laws necessary to establish order and a certain amount of predictability. No one argued on behalf of future generations or the public at large, much less the environment. John Muir, perhaps America's most famous conservationist of the nineteenth century, was only twenty-two when Colorado Territory was created and had yet to find his way to California and the Sierras. And George Perkins Marsh's seminal treatise on the environment, *Man and Nature*, would not be published until 1864. No voices were raised on behalf of what today might be called multiple use or environmental quality, or issues such as fisheries, water quality, or recreation. The concept of sustainability—whether the doctrine could be sustained indefinitely in practice—was not considered. It was "take what you want," unhindered by anything other than what was needed at a particular time and place. Along with water's zero cost to the user, this attitude would control economic development of the American West for decades to come. Claiming land and water, uniquely American, was blessed by the federal government and characterized America's westward migration in the nineteenth century.

There is no surprise here. Humankind's modus operandi had for centuries been predicated on the belief that nature existed for its benefit. In the fourth century B.C., the Greek philosopher Aristotle asserted that "nature has made all things specifically for the sake of man." Two thousand years later, Francis Bacon, the seventeenth-century British philosopher and scientist, declared that "the world is made for man." Given such underpinning, it should not surprise anyone that the American West's first settlers acted with this idea fixed in their minds. *Of course* they would take, indeed should take, what they wanted. If Congress had dictated otherwise, the West might have developed much differently; however, imposing strict requirements at the time was beyond Congressional capacity to govern the West.

Gold continued to draw immigrants to Colorado in the 1860s, though many were drawn less by gold than by the availability of land under the federal 1862 Homestead Act. Under the act, any citizen or citizen-wannabe who met minimal qualifications could claim a hundred sixty acres of unsettled federal land. Following five years of continuous residence, plus payment of a nominal fee, the land was theirs. If the discovery of gold in Colorado was sufficient to draw thousands of immigrants westward, the availability of land made the move even more attractive. Four hundred thousand acres of land were claimed and settled under the Homestead Act between October 1863, when the U.S. Land Office opened in Colorado, and June 1866.[10] Over the following century, more than twenty-two million acres, one-third of the state, would be homesteaded.[11]

Another catalyst that affected the San Luis Valley in the late 1860s was the arrival of Civil War veterans.[12] Most streams draining the mountains that surrounded the valley had already been settled by Spanish-speaking farmers, so the first Euro-American immigrants and war veterans settled primarily in the valley's north end, where they gradually came to dominate that portion of the valley. Irrigated farming was common, but livestock ranching also grew in prominence. While the first two decades of irrigation in the San Luis Valley occurred primarily along the valley's tributary streams, the valley floor was effectively turned over to livestock. Within another decade, the valley changed from a remote collection of Spanish-speaking farmers practicing ancient irrigation methods to a broad desert valley filling with Euro-American homesteaders and ranchers.

In the run-up to the 1876 presidential election, President Ulysses S. Grant wanted Colorado to achieve statehood in order to assist continued Republican control of the federal government. A new state constitution was written and adopted, and Colorado was accepted into the union that year as the thirty-eighth state. Colorado's new constitution specifically adopted and reinforced the Doctrine of Prior Appropriation and added several wrinkles that still form principal components of Colorado water law in the twenty-first century. First, Colorado's water was proclaimed to be the "property of the public...dedicated to the use of the people of the state, subject to appropriation." An important caveat to this proclamation was that the public's role in determining the use of its water ended when the water was appropriated. Second, and strikingly, the constitution states that the right to appropriate state water "shall never be denied." It did not declare "maybe," "come back later," or require compensation be paid to state coffers. Nor did it mandate reserving some water for the public's future use. Likewise, sustainability still did not merit consideration. With these statements, Colorado announced that, from statehood, the public's water resources were free for the taking. There was no defined upper limit, the obvious implication being that water could be taken until there was none left to appropriate. It was the frontier; practically everything was free for the taking. The thought that an upper limit existed, or that a resource might be exhausted, crossed very few minds. Claimants took their place in line until the water was gone, until the rivers ran dry.

Natural streams and rivers were declared a public resource, but the new constitution granted a usufructuary right to use the public's water to anyone meeting the requirements of a diversion and its application to a beneficial use. A *usufruct* is "a right of enjoyment, enabling a holder to derive profit or benefit from property that either is titled to another person or which is held in common ownership, as long as the property is not damaged or destroyed." In effect, the constitution handed the public's water to private parties for personal gain on a first-come, first-served basis. From the very beginning, Colorado allowed and even encouraged private appropriation of a public resource for personal or private gain.

A third wrinkle offered in the new state constitution asserted that "priority of appropriation" had preference when competing water claims required resolution.[13] A recent team of analysts summarized priority this way: "In its strictest form, prior appropriation allows the most senior right holder to divert his water first, the next most senior second, and so on until all rights are met even if it means leaving the stream dry."[14] The early West, including Colorado, was seen as a giant gift box for whoever got there first. In addition to timber and minerals, the gift box included water.

In 1876, such largesse apparently made adequate good sense to those writing Colorado's constitution. Practically everyone in Colorado at the time had direct need for water. Water claimants were, for all practical purposes, the public. The miners and farmers who diverted Colorado's streams and rivers and put them to use in their mines, mills, and on their fields expended considerable physical effort and financial capital. It did not take long for irrigation companies to form and begin excavating major canals, a process requiring considerable capital investment. If the right to water became suspect or was denied, their efforts would be wasted and the capital lost. Labor and capital were at risk and depended upon a continuing availability of free water.

Colorado's grant of private rights to a public resource in 1876 may seem overly generous in hindsight. In the twenty-first century, such largess seems outdated. But at the time of statehood, its intent was, at least in part, to protect economic investments. In this respect, the noted historian Donald Pisani notably concluded "that economic needs and conditions determined the shape of water law more than aridity or any other single physical factor."[15] Colorado's aridity and scarcity of water certainly played a role in forming Colorado water law, but contemporaneous economic factors were extremely important as well.

Today, Colorado has more than five million residents, and water rights, having subsequently become transferable and saleable, are bought and sold like suburban lots. When water rights became private property rights, the public lost considerable input into how the state's water resources would be used. Court decisions and relevant state regulatory agencies would eventually recast and limit this one-sided system, but these elements were not present at statehood.

Colorado's mining industry was born in 1858 with the discovery of gold. Within eighteen years, prospecting and mining were widespread. Concurrently, farmers began plowing virgin ground and raising crops to feed themselves, miners, and other recent arrivals. Denver and the state's burgeoning camps and communities all needed to eat. Colorado's first immigrants wasted no time laying claim to water, which, after soil, is the most important resource necessary for agriculture. Water's scarcity made it as valuable as gold. But where gold had a cash value, water was free. The decisions by which Colorado set aside Riparian Doctrine in favor of a new Prior Appropriation Doctrine would have tremendous implications for the future.[16]

CHAPTER FOUR

AGRICULTURE SETTLES IN

The Rio Grande del Norte heads in the heart of the San Juan Mountains... The agricultural capabilities of this stream and its branches are measured solely by the supply of water... The San Luis Valley is now mainly given over to grazing, but it is beyond question that it will yet be the garden-spot of the State.

— Ferdinand V. Hayden, *The Great West: Its Attractions and Resources* (1880)

Congress encouraged America's westward migration, but it knew it could not expect citizens to pack up their heirlooms and necessities and relocate without knowing something about their destination. To help satisfy this need, after the Civil War ended in 1865 Congress commissioned one of its more farsighted programs, authorizing what became known as the Great Surveys.[1] Beginning in 1867 and continuing through 1878, four unique men led multiple expeditions to acquire hard information about the geology and minerals, water resources, climate, soils, and what did and did not grow in the West. Assisted by geologists, surveyors and topographers, agronomists, botanists, zoologists, artists, and photographers, these scientific leaders catalogued huge expanses of the West. Why men of their caliber pursued such a mission is a reasonable question. Probably they were driven by the same things pushing others at the time: adventure, curiosity, the desire for a new start after the Civil War's debasing cruelties, jobs, human ego, and just maybe a modest serving of Congressional pork. In the end, their reports stimulated interest in the West, lessened the unknowns, and reduced the risks of emigration.

Of the four men, the best remembered in Colorado is Ferdinand V. Hayden, an explorer and geologist who enjoyed a long career spreading his talents over much of the interior West. Before the Civil War, he explored and mapped the geology of what today are the Dakotas, Montana, and Wyoming. Plains Indians named him "Man Who Picks Up Stones Running," alluding to his habit of walking briskly across the prairie and periodically stopping to pick up a rock and examining it before either tossing it aside or putting it in a specimen sack slung over his shoulder. Anyone familiar with the peculiarities of modern field geologists would find this observation spot-on.

Though Hayden's career focused on geology, he was educated as a physician, his stretch as a geologist being interrupted when his medical training was required by the Civil War. Perhaps the most media savvy of Colorado's first serious explorers, he popularized his efforts by routinely publishing magazine articles. His biographer later concluded, "It would be difficult to exaggerate the enormous impact Hayden had on popularizing science in the West."[2]

One of Hayden's broadest and most insightful observations, one with as much currency in the twenty-first century as in the nineteenth, was noting the difference between streams and rivers of the more humid region east of the Great Plains and those in the West. Streams and rivers in the East tend to increase in size and flow in the downstream direction, whereas many western rivers do not. "It is a peculiar feature of these western streams, at times to be larger toward their sources than at their mouths.... Hence all over the Rocky Mountains regions in the autumn are what are called dry creeks, with beds which, when full in the spring time, form large rivers."[3]

Hayden's observation eventually would be explained by fundamental geological and hydrological tenets. Rivers in North America's humid regions increase in flow in the downstream direction because of the progressive addition of tributary and groundwater inflows. By contrast, streams and rivers in the West commonly lack significant tributary contributions precisely because of the region's aridity. Many lesser tributaries in the West's arid and semiarid valleys and plains flow only in response to major precipitation events. Otherwise, they are dry. Aridity also explains why water tables frequently are too deep to provide groundwater to surface streams. The arid climate explains much about the "peculiar feature" described by Hayden.

During a visit to the San Luis Valley in 1868, Hayden observed a feature that would play an extremely important role in the valley's agricultural future, a "Closed Basin" in the valley's north end. He commented on its unique hydrology.

About the center of this park is a singular depression… Into it flow some twelve or fifteen good sized streams, and yet there is no known outlet, neither is there any large body of water visible. It seems to be one vast swamp or bog, with a few small lakes, one of which is said to be three miles in length.[4]

Hayden was also acutely aware of the role timber would play in the area's development and considered its scarcity one of its principal drawbacks. While acknowledging thick pine forests in the surrounding mountains, he bucked the boosters and speculators of the day by denying that the supply was "inexhaustible." He was especially critical of the mining industry. "The rapid increase of the mining operation and population in the mining sections, which are in the heart of the pine regions, is rapidly consuming…the pines around these points. And the numerous fires which occur here, and sweep up the mountain side with a wild fury, like that of a burning prairie, are destroying vast quantities of this timber. Even now we can scarcely travel a single mile along the mountain canons where we do not see the slopes on either side marked by broad strips of burnt timber."[5]

Clearly aware of water's importance to a viable agricultural economy, Hayden concluded by the end of his 1868 visit that irrigation was absolutely necessary. He did not reach this conclusion alone, however. Cyrus Thomas, a noted agronomist of the era, accompanied Hayden on his trips to Colorado and described the San Luis Valley with a sharp eye, noting that streams on the valley's east side generally dried up in late summer, whereas those on the west and north seemed to be perennial. He opined that irrigated agriculture around San Luis Lake in the center of the valley and its Closed Basin would be possible.[6] And while Hayden was highly critical of the Spanish culture taking root in the San Luis Valley, Thomas was more circumspect, commenting that the Culebra Creek near the Spanish settlement of San Luis was "one of the prettiest and richest valleys." Perhaps Thomas's most perceptive observation was directed at the Rio Grande, with his suggestion that a canal might be constructed from where the river leaves the mountains, then extended north and east so it could be used to irrigate an immense portion of the valley floor between the Rio Grande and Saguache Creek on the valley's north end.[7]

By 1876, while political machinations in Denver and Washington, D.C., drove Colorado's pursuit of statehood, farming and ranching in the San Luis Valley were growing briskly. Miners in the surrounding mountains depended on the valley's agricultural productivity by both the Spanish and recent Euro-American settlers. The valley's water resources appeared adequate for anticipated growth, but every farm and ranch required water and their numbers increased every year. The valley's water sources were subsequently documented by Lieutenant George M. Wheeler of the U.S. Army Corps of Engineers, a colleague of Hayden and one of the other three men credited with the Great Surveys. The least recognized of the four in Colorado, Wheeler led major surveying and mapping efforts in what would become Nevada, Arizona, and New Mexico.[8] Between 1874 and 1879, Wheeler led mapping efforts in and around the San Luis Valley. He also recorded his observations of current agricultural activities, including cropped and irrigated fields and water conveyance ditches.[9] The accuracy of Wheeler's maps was a vast improvement on the crude versions in use before the 1870s. Accurate maps were critical to the development of the San Luis Valley, and the streams and river that would underpin its agricultural future were charted to a degree previously unseen. Without reliable maps, settlement of the American West may have flagged because property descriptions would have been difficult to portray. One could walk a property line, but showing an accurate, two-dimensional representation at a desk, on a kitchen table, or in a courtroom would have been impossible without good maps. The siting of roads, canals, bridges, mines, towns, and railroads would have been cruder and taken longer. Without accurate maps, more time and money might have been spent in court than in accomplishing anything on the ground.

A notable characteristic of Wheeler's 1870s maps was the clear and plentiful evidence of how much humans had accomplished in the San Luis Valley since the first Spanish farmers had arrived in 1852. By late 1878, the Denver and Rio Grande Railroad had reached Alamosa. Roads and trails crisscrossed the valley, linking the towns, stagecoach stops, and Rio Grande ferries, as well as passing north over Poncha Pass to the Arkansas River drainage, over Mosca Pass and out of the valley to the east, and south to Taos.

Wheeler provided useful detail about the valley's water resources. Saguache Creek was a much smaller stream than the Rio Grande, yet its irrigation influence was nearly three miles wide near the town of Saguache. Above the town, the creek supported irrigation schemes fifteen miles into the mountains, while below the town Saguache Creek split into distributaries and disappeared into marshes that extended all the way to San Luis Creek in the valley's center.

Conditions differed along the valley's east side, where every note-worthy stream flowing into the valley from the Sangre de Cristo Range was at least partially diverted for irrigation before metamorphos-ing into marshes that merged with San Luis Creek. San Luis Creek's marshes flowed to within several miles of San Luis Lake and its sur-rounding marshes, but no farther. The lake was essentially the terminus for the Closed Basin's streams. South of San Luis Lake, Wheeler's maps indicated mostly "arid and barren" land.

When one appraises the vastness of the San Luis Valley, the snow-pack of the surrounding mountains, and the amount of water dis-charged into the Closed Basin every year, why no perennial surface stream linked the valley's north end with the Rio Grande is puzzling. Gunnison and Beckwith had described the north end's bogs in 1854, and their local guide warned them against trying to cross the valley directly to Saguache Creek because of impassable marshes. Less than two decades later, Hayden described the north end of the valley as "one vast swamp or bog,"[10] which was another reason why early Spanish farmers selected better-drained land along tributary streams at the foot of the mountains.

Wheeler's 1870s maps of the San Luis Valley's south end indicated only a slightly less intricate system of streams and agricultural prac-tices, but the south end's streams joined the river, and all supported irrigation.[11] Creekside bogs and marshes were infrequent. Culebra Creek above and below San Luis was well developed by the 1870s. Small farms, homes, and irrigated fields existed along the creek to within sev-eral miles of the river, while acequias were numerous and prominent.

On the valley's southwestern margin, three significant streams flowed from the San Juan Mountains. Conejos Creek (eventually renamed the Conejos River) is Colorado's largest tributary to the Rio Grande. By the 1870s, irrigated agriculture on the acequia model existed along Conejos Creek from the Rio Grande, upstream to the town of Conejos and into the mountains. In places, fields were irrigated up to two miles on either side of the creek, indicating massive acequia undertakings. Homes and irrigated fields also existed along La Jara and Alamosa Creeks as they flowed toward the river, although as the creeks approached the river their meandering channels, nearly parallel, swung northward and split into multiple waterways before dispersing into what Wheeler labelled the Alamosa Marshes.

In 1880, more than a decade after his initial visit, Hayden penned a description of the San Luis Valley to please the land speculators who entreated his support. "The farming resources of the San Luis Valley are vast and of untold value. In the years to come it will be a beehive of agricultural industry."[12] Like Thomas, Hayden offered an encouraging description of the San Luis Valley's agricultural opportunities and estimated that more than 500,000 acres of land could be supplied with water enough for cultivation.[13]

Settlement and agricultural development of the San Luis Valley occurred briskly over several decades following initial Spanish settlement. Irrigation became widespread around the valley's perimeter but large irrigation projects avoided the Rio Grande, while the valley floor's poor drainage limited its use mostly to livestock grazing. But more changes, more demands on the valley's water resources were imminent.

CHAPTER FIVE

Era of Bonanza Farming

*By the end of the nineteenth century irrigation had become
a veritable crusade, urged on moral, patriotic, religious,
economic, and scientific grounds.*

— Donald Worster, *Rivers of Empire: Water, Aridity,
and the Growth of the American West* (1985)

By the early 1870s, many farmers believed that federal assistance
for irrigation projects was desirable.[1] Given the significant costs
of large irrigation projects, inviting the nation's deepest pockets
to participate made sense to them. As early as 1864, and building upon
western civilization's fixation on dominion over the natural world, the
nascent *Rocky Mountain News* editorialized that the federal government should be involved in irrigating the American West. "Every drop
of water that emerges from the great mountain chains of the West in
their thousands of streams, should be made useful."[2] The definition of
useful, and who decided, was not specified, although it is a reasonable
assumption that the swelling irrigation community would willingly
offer its support and expertise. Irrigation was essential. Without it,
crops could not be grown, miners and railroad workers could not be
fed, and towns could not prosper.

In the San Luis Valley, there appeared to be plenty of water. From
when Spanish farmers began irrigating in the early 1850s, through the
valley's welcoming of Euro-American settlers following the Civil War,
farmers had good reason to assume the water supply was "practically
inexhaustible."[3] Consequently, using it efficiently seemed unnecessary.

Acequias remained in use in the valley's south end, particularly
in Conejos and Costilla Counties. In February 1866, the Territorial
Legislature enacted special legislation authorizing continuation of

the acequia culture, including election and empowerment of a super-intendent, or mayordomo.[4] Neither the approach employed by Spanish farmers nor irrigation techniques utilized by Euro-Americans was efficient, however; irrigation infrastructure differed little between the two cultures. Water was diverted from streams and directed through irrigation ditches to the high end of the fields, where creative means were employed to move water across a field with shovels and pieces of canvas. Called *flood irrigation*, water was allowed to flow overland in natural, shallow depressions or hand-dug or plowed furrows, soaking the soil through the root zone along the way. Soil wetting and satura-tion occurred from the surface downward.

Once a field was wet, water was redirected elsewhere, but two other things had already occurred, one unavoidable because of the nature of the process, the other as a consequence of water's abundance. First, by the time the root zone at the lower end of a field was fully wetted, water still flowing at the top of the field had more than saturated the root zone and continued to percolate downward to the deeper groundwater system. In effect, by the time a field's lower end was irrigated the upper end was over-irrigated. Flood irrigation required more water than the amount simply required to water crops. But water was free and plenti-ful. Farmers could afford to be inefficient.

The second occurrence was that water flowing off the bottom of a field—referred to as *return flow*, or perhaps *wastewater*—was directed away by a drainage ditch or collected in a pond. When possible, this water was captured and reused by a neighbor, giving rise to the saying that "One irrigator's waste is another's supply." Irrigation water not con-sumed by a crop was returned to the local hydrological system, which, depending on multiple factors, could be the near-surface ground-water system, ditches or channels leading to neighboring streams and rivers, or for reuse in nearby irrigated fields. These processes occurred concurrently; their frequency and intensity depended on local condi-tions. Water keeps moving. Overlain on these irrigation practices was Colorado's new water law. Those with senior water rights received all their water before those with more junior rights received any. Junior water rights had to wait for what might be left.

In spite of Colorado's 1876 constitutional commitment to prior appropriation, details of a workable water rights system needed elab-oration. It was not enough for the state constitution to declare that the senior right had priority of use; informal agreements were just as inadequate. A governmental system was needed to oversee and enforce

the law. The Colorado General Assembly undertook to solve the problem with the Adjudication Acts of 1879 and 1881. They remain the nuts
and bolts for establishing legal water rights in Colorado in the twenty-
first century. Every person professing a water right had to file a written
claim in state court to establish the validity of their claimed right. The
claim had to indicate the amount of water diverted, the date of first
diversion, and the beneficial use the diversion served. A judge considered the evidence and, provided that he agreed with the claimant,
affirmed the claim and established the priority date of the right in a
process termed *general adjudication*. Eventually the judge generated
a priority list for each stream under his jurisdiction and delivered it to
state officials, who then administered the rights in accordance with the
list. The judge's declaration of an individual's water right identified the
amount of water that could be diverted and from where, the priority
date, and other key elements, collectively referred to as a *decree*. The
terminology remains unchanged today. Those who refused or failed to
participate in a general adjudication were left off the judge's list and lost
their right to divert and use water as against those who participated
in the proceedings. Historical use was unrewarded if a claimant failed
to participate in the legal process. Lethargy, stubbornness, and anti-
government sentiments were penalized harshly. In effect, "a water right
without a decree has little or no value."[5] This is as true today as it was
in the nineteenth century.

In 1870, the San Luis Valley's entire population was fewer than four
thousand six hundred, representing a substantial influx of settlers in
just two decades.[6] In addition to a few major towns along the river,
lesser settlements were born: Loma, La Loma del Norte, San Jose,
Venable, Fish, Williams, and Spalding. Of these, the first four indicate
origins associated with Spanish settlers. The latter three suggest later
Euro-American settlers. Irrigated agriculture flourished on the valley's
perimeter, where streams were linked to the river by a physical channel or through distributaries and marshes. Irrigated agriculture was
gradually increasing along the river between Del Norte and where the
river began to incise through the San Luis Hills, a distance approaching
one hundred miles and evidence that the valley's newest farmers were
willing to test their skills and patience against a river that flooded every
spring. The river meandered, sometimes in multiple channels, across a
mile-wide floodplain.

Ferdinand Hayden's predictions about the San Luis Valley's agricultural future came true almost as fast as he pronounced them. The

climate, though cool, was sundrenched, and the valley's fundamental nature was desert; away from the river, streams, and marshes the land was dry. After the best farmland along the river and the valley's tributaries was settled and put under irrigation, latecomers were pushed to the margins, where the productivity of their farms and ranches was limited by scarce water. The Rio Grande was far from being totally tapped, however, and the valley floor certainly was flat enough for irrigation, but there was no way to move river water to the valley floor. Consequently, the valley floor was used primarily for livestock grazing.

In 1872, the agronomist Cyrus Thomas had suggested that a canal could be constructed extending north and east from where the Rio Grande left the mountains on the valley's west side. Thomas postulated that such a canal and connecting laterals and ditches could irrigate much of the valley floor between the river and Saguache Creek on the valley's north end.[7] By 1879, one hundred twenty-two thousand acres, nearly two hundred square miles, were already being irrigated in the valley, but tens of thousands of new acres could be opened to irrigation if such a canal were built.[8]

Theodore C. Henry, one of the valley's early settlers and possessing a speculative spark, persuaded The Traveler's Insurance Company, based in Minnesota, to lend money to the spanking new Del Norte Land and Canal Company to buy un-irrigated land in the valley. It was speculation on a grand scale, a phenomenon that would be repeated throughout the nineteenth-century West. Following this purchase, Henry organized the financing and construction of four major irrigation canals leading from the river to the purchased land. Henry's efforts included the Rio Grande and San Luis Valley Canals, which extended north of the river, and the Citizen's Ditch (later renamed the Monte Vista Canal) and Empire Canal south of the river. Numerous existing ditch systems were incorporated into the canal network. The four canals totaled one hundred forty miles in length and eventually supplied water to a complex of laterals capable of irrigating five hundred thousand acres.[9] Once water was available the land became substantially more valuable.

The Rio Grande Canal was by far the largest waterway. Begun in 1881, it took three years to complete and required, varying with the season, the efforts of a thousand teams of horses and between three and five thousand men. Its engineered design flow capacity was 2,540 cubic feet per second, or CFS.[10] At its headgate on the river, the canal was sixty feet wide at the bottom, ninety feet wide at the top, and six feet deep. Running full, it could divert 30 percent of the river's average annual flow.[11] The combined capacity of the Rio Grande and the first three additional large canals, when they all came online in 1889, was 3500 CFS. Compared to a typical summer flow in the river at Del Norte of about 1500 CFS, these canals were capable of dewatering the river during all but peak spring flows.

By modern standards, the irrigation system resulting from these canals was fundamentally wasteful of water. Efficiency was ignored. One trio of historians lamented that, "Even if the water was not needed for irrigation in the San Luis Valley, it was diverted anyway. The unneeded water was dumped down the local creeks where some of it overflowed to irrigate meadows. The residual ultimately drained into the Closed Basin."[12] This surfeit of water triggered another wave of irrigation and settlement in the San Luis Valley. Lasting into the early 1890s, this period of rapid irrigation growth became known as the era of "bonanza farming." Altogether, more than four hundred thousand acres of new land came under irrigation, effectively tripling the amount of irrigated land in little more than a decade. The valley's major crops were wheat, oats, barley, field peas, alfalfa, and potatoes.[13] All required irrigation, especially alfalfa, an especially thirsty legume used as a high-value hay crop. Water was the *lingua franca*. Everyone spoke and understood it. Farmers and ranchers worked exceptionally hard and paid for the privilege with shortened lives: the average life expectancy in the valley was just forty-seven years.

Farming fed farm owners and their families, nearby communities, and miners in the surrounding mountains. At statehood in 1876, the San Luis Valley had a population of less than five thousand. The valley's major communities—Alamosa, Monte Vista, and Saguache—grew to a few thousand residents each, yet the valley's population density in 1900 would still be less than three persons per square mile.[14] No one would have suggested that the San Luis Valley was anything but rural farming and ranching country. Domestic and municipal water needs and consumption, compared to irrigation water use, were inconsequential.

Transportation remained critical to the valley's growing agricultural economy, as well as to mining activity in the San Juan Mountains to the west. One important component of this need was met by railroads. The Denver and Rio Grande made it to Alamosa by 1878, after which the river was no longer a major impediment to travel and the transportation of goods and products.[15] By 1891 a new mining boom required extending the rail line to Creede. Nor was the valley's north end ignored. Tracks crept over Poncha Pass in 1881; tracks joined the main line between La Veta Pass and Alamosa by 1890. The Denver and Rio Grande Railroad linked the valley's agricultural and mining centers and integrated them with Colorado's Front Range.[16]

By 1889, when all four large canals were functioning, the Rio Grande was frequently dewatered during the growing season. Little consideration was afforded Spanish farmers downriver in the valley's south end, most of whom had complied with the new prior-appropriation water law. Similarly, no concern was afforded New Mexico Territory or Texas. No agreement existed among the states and Mexico on how to apportion or regulate their interests in the Rio Grande. Being the headwaters state, Colorado did as it pleased, a routine that lasted through the remainder of the nineteenth century.

The era of bonanza farming, due almost entirely to irrigation, led to the limited availability and increased cost of irrigated land. Some innovative individuals responded by attempting to exploit the valley's groundwater. Isolated shallow wells in the valley's near-surface *unconfined aquifer* had already been completed by those wanting domestic water for homes and gardens, but development of the valley's groundwater for irrigation did not begin in earnest until 1887, when the first well was completed in a deeper, *confined aquifer*.[17] Within the next four years, water from an estimated two thousand wells in the valley literally flowed at the surface, some at hundreds of gallons per minute.[18] Pumps were unneeded.

Initially, farmers were suspicious and untrusting of groundwater. Groundwater is invisible. Its occurrence can be inferred but proven only by drilling or digging. Conditions defining its existence in the San Luis Valley were unknown in the nineteenth century because geologists had yet to thoroughly characterize the valley. Geology as a science was still young, groundwater science and hydraulics even younger. Groundwater eventually would become heavily used for irrigation, but that time was still decades away.

Colorado's depletion of the river did not go unnoticed in New Mexico Territory, Texas, and even Mexico. By 1889, the Rio Grande was frequently dewatered before leaving Colorado,[19] and although tributary inflows farther south restored some of the losses to downriver users, one historian of the river and its irrigation practices reported that water use in Colorado and New Mexico resulted in shortages as far downriver as El Paso and Juarez, Mexico.[20] After drought took hold in the region in the early 1890s, Texas and New Mexico Territory complained to the federal government. In time, their complaints were heard. The International Boundary and Water Commission, a predecessor to the U.S. Reclamation Service and subsequently the U.S. Bureau of Reclamation, undertook an investigation of water diversions from the Rio Grande between its headwaters in Colorado's San Juan Mountains and El Paso, Texas. Completed in 1896, the investigation concluded that, drought or no drought, downriver water shortages were indeed due to major canal construction in the San Luis Valley in the 1880s.[21] No one in Colorado cared, however. Colorado water law provided reliable and predictable access to a limited resource that was free for the taking. Just grit, sweat, and determination were needed, something Americans, particularly those on the western frontier, possessed in abundance.

In less than a half-century, the upper Rio Grande and San Luis Valley had undergone tremendous change. Native Utes had been evicted from the valley, allowing Spanish settlers to migrate up the Rio Grande from New Mexico Territory and put down roots. A decade later, prospectors and miners arrived, lured to Colorado by gold, joined by Euro-American settlers pouring west after the Civil War to take advantage of the 1862 Homestead Act. The Rio Grande had been tapped by canals such that the river was frequently dewatered above the state line for days or weeks at a time. From a population of less than four thousand six hundred in 1870, the valley had grown to nearly eighteen thousand by 1890.[22] The San Luis Valley had been transformed from a high desert valley of incongruous, impassable marshes, disappearing streams, and barren ridges into an agricultural bonanza. Hundreds of thousands of acres of irrigated farms had been developed.

By the 1890s, the Rio Grande was no longer the river it had been less than a half-century before. A limit had been reached. So much change had come to the valley and the river in such a short period of time that someone would have been justified in asking whether there was any capacity for moderation. More likely was a stone-cold belief that everything would be just fine if downriver neighbors simply stopped complaining.

THE ROLE AND IMPORTANCE OF STORAGE

You can't take water from a 90-day runoff and apply it for the
rest of the year without storage.

— Steve Vandiver (2013)

Т he road to the Rio Grande Reservoir is rough and dusty and
begins only after driving for several hours out of the San Luis
Valley, past South Fork and Wagon Wheel Gap, through Creede
and beyond, climbing toward the Continental Divide. Below Spring
Creek Pass a gravel track leaves the pavement, winds across a grassy
bench, crosses ridges, circles ponds and small lakes, and switchbacks
out of dead-end canyons, picking its way as if seeking the reservoir for
the first time. The Rio Grande finally appears a thousand feet below,
twisting across a narrow valley floor at the bottom of a sunny slope
spotted with blue-green spruce. The river meanders, then splits. Both
channels straighten, rejoin, and meander again until the river disap-
pears into up-valley shadows. The valley's opposite slope, north-facing
and partially hidden from the sun, supports a handsome pine-and-
spruce forest. Midway up the slope, it is broken by wine-colored cliffs.
Above the cliffs, the forest reappears and the slope merges into the
shoulder of a mountain arched against the horizon.

The road descends the slope and parallels the river, its icy water
palpable through the open truck window. The floodplain is a willow
jungle. Continuing, the road passes two Forest Service campgrounds
and rises skyward one final time before a dam, a giant plug in the valley,
appears. The river foams from the dam's base. Behind the dam, a cobalt
blue reservoir stretches deeper into the mountains.

Travis Smith stands on the dam crest, waiting, a ready smile on a
face partially hidden behind sunglasses. He is six foot three and stands

a head above everyone nearby. Smith has been superintendent of the San Luis Valley Irrigation District for more than twenty years and knows as much as anyone about the upper Rio Grande and San Luis Valley farming and irrigation. The district owns the dam beneath his feet, the Rio Grande Reservoir, and the valley's Farmers Union Canal.

Today, Smith is meeting federal officials to discuss a dam access road realignment that may require Forest Service approval. The occasion also warrants a neighborly invitation to the county commissioners. The dam and reservoir are on private land in Hinsdale County that is surrounded by national forest. He does not expect problems with the meeting; it is to inform and explain, to set an agenda and manage expectations for future negotiations, if needed. He is articulate and sincere, talents acquired from years of conversing with curious citizens and farmers in need of water. Smith speaks the language of water fluently.

The dam crest beneath Smith's feet stands 110 feet above the valley floor and stretches 550 feet across the canyon. Far from the San Luis Valley and deep in the mountains, the dam and reservoir are isolated. A caretaker lives on-site during the irrigation season in a house provided by the district. The caretaker's job is to control the reservoir's outflow in accordance with instructions telephoned by Smith, who normally spends his days in the valley. Smith and the caretaker juggle as many as eight different requests, or "calls," for water from the reservoir each day.[1] In addition to storing water for the district, the reservoir also holds water for other parties, including the Colorado Department of Parks and Wildlife. These other needs and associated water rights are not the same as those of the irrigation district, but all demand frequent adjustments to reservoir releases.

Before he took this job, Smith was the water commissioner for District 20, a geography that includes the main stem of the Rio Grande in the San Luis Valley. Colorado's water commissioners control the headgates that divert water from the state's streams and rivers. They are the bookkeepers and ensure that the correct amount of water is diverted. They track the river flows and balance them against decrees specifying which water rights are "in priority" and thus eligible for water—and which are not and must remain unfilled. To some farmers, the water commissioner is the closest thing to a god they may see during the growing season. To a few, the commissioner is a pain in the ass because he correctly turns their water on and off according to decrees. In the process, he may make their hard days harder. Water commissioners carry both the burden of responsibility and the regulatory mallet of the state and division engineers. If a water user does not play by the rules set by law and prescribed by the division engineer, the commissioner can close and lock the malcontent's headgate. This is a major deterrent, obviously, and one infrequently needed.

When the irrigation district's water is released from the reservoir, the river conveys it to the Farmers Union Canal headgate below Del Norte, downriver and miles away. The water's arrival time depends upon constantly changing weather, additions to and other diversions from the river, and the varying flows and hydraulic character of the river. The time between the releases of water from the reservoir and its arrival at the district headgate can be measured in multiples of hours or even days. Irrigators must plan ahead.

A commissioner's responsibility ends after he or she has diverted water from the stream or river. From that point, a "ditch rider" working for a district, farm, or another canal company shepherds the water through more headgates and ditches to the valley's hundreds of irrigated fields. Individual farmers with a relatively small number of irrigated acres may manage their own headgates, while the valley's large canal companies employ fulltime ditch riders whose responsibility is to ensure that the right amount of water gets through the valley's labyrinth of canals, ditches, headgates, culverts, and ponds. It is a complex system based on the repeated division of moving water. It depends on accurate recordkeeping and knowledgeable workers, all performing defined tasks in known and repeatable ways. The valley's farmers obsess about this reliability. Their livelihood depends on it.

The San Luis Valley Irrigation District evolved from the Farmers Union Irrigation Company, which owned and operated the Farmers Union Canal,[2] one of the valley's major canals, constructed in the 1880s. With a decreed capacity of just over eight hundred CFS, it is the smallest of the valley's six largest canals.[3] Its water rights are junior to those of the five larger canals, as well as several hundred smaller ditches along the river, all dating to the 1880s or earlier. During its first few decades, the Farmers Union Canal frequently had its water turned off during droughts and when late-summer river flows could not satisfy all decreed water rights. Water stored in a reservoir would help alleviate late-season shortages; consequently, the Rio Grande Reservoir was planned. The plans were complicated, however, when the International

Border and Water Commission's report was released in 1896,[4] blaming downriver water shortages in New Mexico Territory and Texas on the San Luis Valley's canal-building craze of the 1880s.

During the drought of the 1890s, Colorado, New Mexico Territory, Texas, Mexico, and the federal government swapped insistent claims over the Rio Grande and how or whether it should be managed. New Mexico Territory and Texas wanted federal intervention, and Mexico threatened legal actions against the United States if it did not get a greater share of the river. Mexico had been using the river for irrigation and domestic purposes since the seventeenth century, long before the San Luis Valley was settled. Being without access to irrigation water was unacceptable in the extreme.

In Colorado it was different. Between 1889 and 1899, in spite of the drought, the number of acres irrigated from the Rio Grande and its tributaries doubled to nearly 296,000 acres.[5] To suggest that a senior water right holder might allow water to continue downriver in a time of shortages to benefit other users was laughable and easily ignored. Why should hardworking, prosperous Colorado farmers care about Texans hundreds of miles distant? Coloradans believed it was just fine to ignore downriver demands. Colorado benefitted immensely from being a headwaters state; it was not dependent upon the generosity or fairness of other states and was beholden to no other entity for its water.

But the interests of San Luis Valley farmers were challenged nonetheless. The commission's 1896 report carried considerable weight in the nation's capital. To pacify Mexico, Texas, and New Mexico Territory, the Secretary of the Interior placed an embargo on the use of federal land in Colorado's portion of the Rio Grande watershed for construction of any new reservoirs. To the federal government and downriver water users, too many irrigation diversions had been constructed, too much water was being withdrawn from the river, especially during droughts. Colorado irrigators were unhappy with the embargo, but the reasoning behind it was clear: new reservoirs in the upper watershed would exacerbate the damages and threats to current and future downriver uses. It soured public attitudes that the embargo "was issued expressly to prevent further depletion of the flow of [the] Rio Grande in the Elephant Butte-Fort Quitman section" along the river in New Mexico Territory.[6] The embargo specifically was to protect the viability of a proposed dam and reservoir project at Elephant Butte, located on the Rio Grande in central New Mexico Territory and nearly three hundred miles downriver.[7] In addition to wondering how Texas and

New Mexico Territory acquired such political clout in Washington, D.C., Coloradans argued that Rio Grande tributaries between the San Luis Valley and Mexico could surely make up the difference. However, analysis of these tributaries failed to work in Colorado's favor. The embargo held. To this day, the embargo remains a sore point for valley residents with long family ties to the river.

The political war over water began to defrost with passage of the federal Reclamation Act of 1902, which granted the federal government authority over planning, financing, and constructing new dams in the West, in lieu of handing these responsibilities to the states or private interests. States' rights advocates, normally averse to a strong federal role in water management, went along because Congress promised money and robust state involvement in nominating projects. The act also stipulated that water for new developments would be governed by state, not federal law. Importantly, the authority to define and create rights to the nation's streams and rivers, including the Rio Grande and its water, was ceded to the states and territories.[8]

In 1904, a tentative settlement was reached between the United States and Mexico that acknowledged the eventual construction of the Elephant Butte Dam and Reservoir. This temporary agreement was finalized with the Rio Grande Convention of 1906, which called for more than 150,000 acres to be irrigated from the river in Texas and New Mexico Territory, plus the annual delivery of 60,000 AF to Mexico. The project envisioned using 730,000 AF for irrigation below the Elephant Butte Dam. Water for this massive irrigation project would be stored in the Elephant Butte Reservoir, an immense 2.1 *million* AF impoundment. Colorado and San Luis Valley farmers felt cheated. The Elephant Butte Reservoir became, in one historian's words, "the tail that wagged the dog" for all water storage projects in the Rio Grande headwaters.[9] Farmers in the San Luis Valley were also angry because, despite the rosy assurances of the Reclamation Act of 1902, they could no longer expect federal assistance or cooperation to build storage in the Rio Grande's headwaters. They had little recourse but to continue dealing with the river's mercurial fluctuations on their own. Worse, the embargo was not immediately lifted.

By the time Colorado completed its 1903 general adjudication of water rights, most of the Rio Grande's flow in the San Luis Valley was being diverted most of the time, all destined for agriculture.[10] More storage was avidly sought, but the Farmers Union Irrigation Company's plans to construct a reservoir in the river's headwaters were stymied.

The valley's prayers were finally answered in 1907 when the federal embargo was relaxed. Ideas about more reservoirs had been kicked around the valley for years, but their construction had always been prevented by the embargo. When the embargo was relaxed, the Farmers Union Irrigation Company quickly reorganized as the San Luis Valley Irrigation District. An advantage to being organized under Colorado's 1905 irrigation district law was that it provided districts with taxing and bonding authority. The new district promptly issued bonds for just over a half-million dollars. Following several years of engineering and site preparation, the Rio Grande Dam was begun. By 1912 water could be stored and made available to district farmers in the valley. The dam and reservoir were completed two years later. It was a typical dam of the era, more earth and rock than concrete. It required hundreds of men and up to a hundred teams of horses to move enormous volumes of earth and rock. When completed, the 54,000 AF Rio Grande Reservoir provided water storage to aid in irrigating sixty-two thousand acres of prime farmland in the center of the San Luis Valley.[11]

Seven additional earthen dams and reservoirs were built in and around the San Luis Valley between 1907 and 1914. All were privately financed.[12] The Santa Maria Reservoir, with a capacity of nearly 44,000 AF, was controlled by the same company that owned the valley's largest canal, the Rio Grande. The Sanchez Reservoir on upper Culebra Creek, the valley's largest private project and with a capacity of over 103,000 AF, was completed in 1911.[13] The eight reservoirs had a combined storage capacity of more than 250,000 AF, just one-eighth of the storage of the proposed Elephant Butte Reservoir in New Mexico. And where this new headwater storage assisted in irrigating 400,000 acres in the San Luis Valley, the Elephant Butte Reservoir, with its 2.1 million AF of stored water, supported fewer acres, although part of the additional storage at Elephant Butte was to provide the 60,000 AF of water previously promised to Mexico. For the San Luis Valley, it was better than nothing, yet few were satisfied.

After San Luis Valley farmers constructed their eight dams and reservoirs, Texas and New Mexico again objected. The Secretary of Interior did not re-impose an embargo, but more reservoirs in the upper Rio Grande watershed once again became little more than dreams. The embargo's relaxation had been enough for valley farmers to make the most of the situation, and the downriver states and Mexico nevertheless settled temporarily for the status quo. In reality, New Mexico was a winner in the interstate water sweepstakes. New Mexico gained

statehood in 1912, followed four years later by completion of the Elephant Butte Dam and Reservoir, resulting in a substantial increase in
storage and water availability in the lower Rio Grande. Only modest
storage has been added to the upper Rio Grande watershed since its
burst of dam-building between 1907 and 1914.

Dams and reservoirs are pivotal to agriculture in the American
West. Without them, it is arguable whether conventional agriculture could thrive in the region. In the San Luis Valley, the reasons are
evident. The Rio Grande moves most of its annual flow in just three
months, while crops require water over a growing season of four to five
months. Without storage, irrigation and farming in the San Luis Valley
would have been handicapped in the nineteenth century.

CHAPTER SEVEN

Making an Aquifer

And while many center pivot irrigation systems are supplied only from groundwater, the practice of artificially recharging the unconfined aquifer with surface water is what sustains the groundwater supply in many parts of the valley.

— William A. Paddock, *Introduction to Water Resources Issues in Water Division No. 3, The Rio Grande Basin* (2014)

North of Del Norte, Colorado Highway 112 crosses the Rio Grande, breaks free from cottonwoods growing on the floodplain, climbs onto a low bench, and bends eastward toward the valley center. Fifty miles away, the Sangre de Cristo Mountains are the horizon. A powder-blue bowl of sky hangs above a valley floor strewn with irrigated farms. A geologist's eye, however, sees not just farms or horizons, but the valley itself as a vast structural basin imposed on the earth's crust. The San Luis Valley is underlain by thousands of feet of clay and silt and sand, gravel, ancient lakebeds, and volcanic ash. Rain, melting snow, and streams have infiltrated and filled the pores in these sediments. The San Luis Valley is a huge tub of groundwater.

A road on the left turns back to the river's north bank and the Rio Grande Canal headgate. In the opposite direction, the highway parallels the canal for several miles before the canal turns north and contours the valley floor's western edge. The canal is the granddaddy of the valley's major irrigation canals. Its water right is among the most senior on the upper Rio Grande, which means that many other ditches and canals, all smaller, receive no water until the Rio Grande Canal receives its full decree. The Rio Grande Canal, along with the six other major canals constructed during and since the 1880s, divert water from the river over an eighteen-mile reach below Del Norte. Five of these seven—Rio

Grande, San Luis Valley, Billings Ditch, Prairie Ditch, and the Farmers Union Canal—spread water to the north, onto the river's broad alluvial fan and toward the Closed Basin. The remaining two canals—Empire and Monte Vista—direct water to farms south of the river. Together, these seven canals were responsible for the Rio Grande's controversial de-watering in the 1880s and 1890s.

The highway continues eastward for six miles until it intersects U.S. Highway 285, which connects Monte Vista with Saguache on the valley's west side. On the valley's eastern edge, Colorado Highway 17 links Alamosa with Poncha Pass and the valley's north exit. Both highways are straight as a rifle barrel for many miles. In the center of the valley is Center, a farming community of several thousand. Highway 112 glances Center's southern edge while the Farmers Union Canal splits it diagonally southwest to northeast. Center's streets, some paved, most gravel, aim at the cardinal directions. Cottonwoods and a few verdant spruce trees grow next to wet spots or near houses. For a town its size, Center appears to have a disproportionate number of churches. Several apartment houses shelter seasonal labor. Private homes are the smallest buildings in town, modest and single-story, and interspersed among a lesser number of modular structures and skirted mobile homes. A Family Dollar, numerous bars, and automotive and truck parts stores constitute the town's business district.

Center appears to be home to more trucks, at least per capita, than anywhere else in Colorado. Cars are rare. Trucks come in all sizes, shapes, and ages, from small, rusting Ford Rangers to six-passenger, one-ton, four-wheel-drive units with enough clearance to pass over ditches and through wet fields, to even larger trucks that haul potatoes from the surrounding farms to Center's rows of storage and packing sheds and railroad depot. Every truck boasts proof of farm use: dents, scratches, cracked windscreens, wind-blasted and sun-faded paint. They are mud-caked and covered with dust. License plates are illegible beneath splatters of mud. From dawn to dusk, trucks rumble down the streets or idle, with the distinctive clatter of diesel engines, in front of the few cafes. Center's gas stations and fuel depots sell more diesel than gasoline. At the town's edge, a jungle-gym of steel farm machinery is parked helter-skelter next to the sheds. Most are for planting and harvesting potatoes, yet are so specialized that a visitor might never recognize their purpose.

Life in Center feels slow but deliberate and reflects the habits of the valley's farmers. Routines run in accordance with the season and the weather. Nature punches the only clock that matters. Yesterday's rain followed by light evening snow has left the fields and roads wet. It is April; irrigation season begins soon. It is only a matter of days until the valley's irrigation canals, laterals, and ditches surge with water and the valley's center pivots begin their measured spins.

Travis Smith is tall, even while sitting hatless behind his desk. His office, shared with his office manager, is spacious, warm and dry, a welcome refuge from a chill morning. Papers and reports cover his desk. The visitor's chair sits beside his desk, not opposite it, and he swivels ninety degrees to face his callers. The office building has been described by Smith and others as an upgraded chicken coop. In the center of Center, in the center of the valley, the San Luis Valley Irrigation District office occupies a metal-clad Quonset hut like those developed in World War II and widely used since on poultry farms. This structure would never be mistaken for anything but an office; a district sign makes it clear. The building occupies a corner lot close to the town water tower. A shed roof hangs over its entrance. In farm country, discussions frequently extend beyond bars and neighboring fence lines to hours hunched over parked trucks or when meeting at the district office. In rain or snow, a roof encourages discussion.

The San Luis Valley Irrigation District is the only formal irrigation district in the valley. Other irrigation enterprises in the valley, like many in Colorado, are organized under state law as mutual ditch companies. Although differences exist in how they are established and operate, irrigation districts and mutual ditch companies, in effect, "own" the consolidated water rights of everyone who obtains their water from a company-owned ditch or canal. Individuals own shares in the district or company proportional to the amount of water they receive. Shareholders typically may sell or transfer their shares only to other shareholders in the same district or company.

Smith's district has about a hundred fifty members who irrigate sixty-two thousand acres, or nearly one hundred square miles of the valley's best agricultural ground and the essential heart of the Closed Basin.[1] Many farmers are fourth and fifth generation residents whose surnames show up on valley maps as corners, ditches, and lanes. Smith, however, is not native to the valley. His parents moved there in 1957, purchased an old homestead, and farmed. After high school and a stint in the Navy, he returned to the valley, attended Adams State College in Alamosa, married a valley-born woman, and stayed.

Twenty minutes of light banter passes before a serious question is broached. What about the valley's water problems? He responds in a single, breathless burst, summarizing the valley's first forty years of irrigated farming. "It's land development, it's the Homestead Act, it's the federal agenda about populating the West." He names the seven major canals in the valley and, referring to the 1890s, concludes, "And we dewatered the river."[2]

Skip ahead a century and things seem to have changed little. Beginning in 2002, a drought has gripped the San Luis Valley. Indeed, much of southern Colorado experienced below-average precipitation, especially winter snowfall in the mountains that fed the reservoirs, rivers, and the valley's irrigation systems. No irrigator will ever admit to having too much water, but Smith believes that the valley's recent difficulties with water shortages began with the advent of high-capacity pumps and wells. In contrast to low-yielding wells used primarily for domestic purposes, which may yield two to twenty gallons per minute, a high capacity irrigation well can produce several hundred to several thousand gallons per minute.

Farmers began to significantly use wells for irrigation in the 1940s. A drought in the 1950s reduced river and stream runoff available for surface irrigation, and well development expanded across the valley. Well construction really took off in the 1960s. The Colorado state engineer attempted to place a moratorium on new well development in 1971, but innovative farmers found ways around the moratorium by drilling "supplemental" wells and "alternative points of diversion," all of which were legal and associated with decreed water rights. Smith says, almost sadly, "Now we're twenty years later [and] we've over-pumped. We're mining [the groundwater]. We've mined 1.2 million AF from the 1976 level."[3]

◆ ◆ ◆

Hundreds of thousands of acres of the San Luis Valley floor are prime farmland. The number varies, but generally fixes on about five hundred thousand acres—nearly eight hundred square miles—more if one includes irrigated acres fed by tributary streams rimming the valley floor. Not every acre is irrigated and farmed every year, but most are. The number varies with family and owner preferences, commodity prices—and available irrigation water. Potatoes, hay, alfalfa, and barley are the principal crops.

The valley's county roads reflect surveying conventions of the nineteenth century. The land is laid out in one-mile squares, or sections. Each section contains six hundred and forty acres; thirty-six sections create a square six-by-six township. Most sections contain neat farms and single-story farmhouses set between modest barns and equipment sheds. Occasional rows of cottonwoods and junipers temper the valley's strong spring winds. Tractors, overhauled and repaired over the long winter, are positioned by the fields, ready for duty.

The most prominent characteristic of this heavily farmed valley floor is the center pivot irrigation system. Center pivot irrigation was invented in 1949 by Frank Zybach, a farmer from the semiarid plains east of Denver. Today, it seems as if every hundred-and-sixty-acre quarter-section in the center of the valley has a center pivot, most of which date since the 1970s. From the quarter-section's center, an aluminum pipe rises from the earth, bends and extends horizontally outward, arm-like, for a quarter mile. Water from the farmer's well or one of the valley's myriad of canals and ditches is pumped through the pipe to sprinkler heads that hang from the pipe, like teats from an udder. As the pipe-mounted sprinkler heads pivot around the center pipe, it traces a circle that is tangent to the four sides of the quarter-sections. "The system could ride over sandy hillocks, requiring no land leveling or ditch digging, throwing water over a field like a light rain falling from the sky."[4] In its simplest mode of operation, the farmer flips a switch, an electric pump turns on, and water pulses through the pipes and sprinkles the crop from the slow-moving sprinkler heads. Irrigated quarter-sections abut one another for miles across the valley floor. From afar, the valley appears like a maze of tall fences designed to contain exceptionally tall horses.

From above, the valley floor appears as a biscuit board of irrigated circles, a circle-within-a-square geometry that leaves the four corners of the fields un-irrigated. Farmers use these dry corners for equipment sheds and storage, irrigation infrastructure, water storage ponds, and their homes. Curious non-farmers sometimes wonder whether these unfarmed corners reduce total farm productivity when compared to a completely irrigated square field. The answer is that overhead sprinkler irrigation is more efficient and more productive than traditional flood or furrow irrigation. Leaving out the corners is economically equivocal. A skilled farmer can grow and earn more from a well-irrigated circle than from a less-efficiently irrigated square. Over the past century, the growth of efficient irrigation—specifically center pivot irrigation and the water distribution system that supports it—have become both a blessing and a curse to the valley.

By 1917, when America entered World War I, most of the San Luis Valley's best accessible farmland had been claimed or purchased, settled, and farmed. Much of it was along the Rio Grande or on the valley's periphery, where mountain tributaries provided irrigation water for the natural hay meadows, called *vegas*, at the canyon mouths. Also included were large swaths of land below the major irrigation canals in the valley's Closed Basin. What these thousands of acres had in common was ready access to water.

A historic drought ended in 1896, but its cessation triggered a new problem that no one had anticipated. Over the next few decades water surplus created a problem as worrisome as water scarcity. How could there be too much water in a desert? In part, it resulted from an irrigation practice begun in the 1880s—still called "bringing up the sub" throughout the West—that entailed artificially raising the water table to within several feet of the ground surface. Given the local habit of diverting the river into the valley whether irrigation water was needed or not, it is a reasonable conclusion that the new practice may have come about indirectly. In some places the water table actually rose to the ground surface. A near-surface water table allows plants to pull groundwater into the root zone by the physical process of capillary action.

As a consequence, irrigation in the San Luis Valley was literally turned upside down. Crops were watered from below, by *subirrigation*, rather than from the surface by traditional flooding. A geologist who spent years documenting the valley's water resources commented on both a major advantage and disadvantage to the new practice. "This method requires much less care and trouble than the method of flooding or surface irrigation, and is as efficacious as that method, though it requires much more water."[5] Subirrigation gradually replaced flood irrigation until its use was all but universal in the valley center where the water table was closest to the surface. Water was substituted for labor, even though the new method used water inefficiently. The substitution of cheap water for the labor required to flood irrigate fields, or the capital necessary to line leaky canals and ditches, was a deciding factor.

The large canals constructed during the 1880s contributed significantly to these changes. The return to normal river and stream flows after 1896 provided more water than the farming community had become accustomed to. But not all water transported by the canals was consumed by crops by means of *evapotranspiration*.[6] Excess water ran off irrigated fields and infiltrated the ground surface, or flowed into sloughs, streams, and excavated surface drains. The primary canals, laterals, and ditches were (and remain) unlined. Water seeped through their bottoms and joined the underlying shallow groundwater system. These processes collectively contributed to a rising water table in broad areas of the valley. Prior to the 1880s, the only portions of the valley floor that were reliably wet were along seasonal tributary streams where they debouched from the mountains and along Saguache and San Luis Creeks.

Because so much water was diverted from the river to the valley floor, because the valley's canals, laterals, and ditches were unlined, and because these factors allowed *bringing up the sub* to become the dominant form of irrigation, a new, unconfined *aquifer* was created beneath much of the western valley floor and extending eastward into the Closed Basin.[7] Intentionally or otherwise, development of irrigation canals and distributary laterals since the 1880s created one of the first manmade aquifers in the western United States.

Few at the time thought about the long-term ramifications of raising the groundwater table between forty and one hundred feet on the valley's west side and close to the ground surface in the lowest areas of the Closed Basin.[8] *Bringing up the sub* might have been appropriate in some parts of the West, but the center of the San Luis Valley was not one of them. Natural salts—the fluorescent salts that colored the earth's surface white, described by Gunnison and Beckwith fifty years before—created saline soils and sediments in the valley's low areas long before humans arrived. When these salts were mobilized by subirrigation and a rising water table, then drawn into the root zone or concentrated at the soil surface by evaporation, the resulting salt concentrations became toxic to plant life. Increased salt concentrations eventually *waterlogged* affected soils, or rendered them so saline or alkaline that crops failed.[9] Yields plummeted. Diverting and applying so much water to the valley floor gradually impaired its soils. The few plants unaffected by these altered soils were primarily native greasewood, rabbitbrush, and other uneconomic flora. Within several decades, as the full cause and extent of soil salinization became apparent, portions of the valley were abandoned by farmers. The most affected areas were near the community of Center and along the valley's east side and paralleling San Luis Creek.[10]

The practice of subirrigation gradually made farmers aware of a previously overlooked source of water: the ground beneath their feet. As this awareness spread, thousands of groundwater wells were completed in the valley after the late 1880s. Completed in a deeper *confined* aquifer, many of these wells were *artesian* and flowed at the ground surface. An estimated 2,000 *flowing* wells existed in the valley in 1891,[11] a number that increased to more than 3,200 by 1904, an estimated 5,000 twelve years later, and as many as 7,500 by 1958.[12] The first flowing wells were used primarily for livestock and domestic needs, and only occasionally to supplement flood irrigation. The confined aquifer did not appreciably contribute to the expansion of irrigation in the valley. That distinction belonged to the shallow, unconfined aquifer created by irrigation and the major canal systems. The first irrigation well completed in the unconfined aquifer came about in 1903, although valley farmers did not begin to seriously exploit that aquifer for irrigation until the notorious drought years of the 1930s, when low river and stream flows drastically reduced the amount of available surface water. Farmers were encouraged to drill wells in the shallow aquifer, and by the early 1950s wells in the unconfined aquifer numbered approximately thirteen hundred. They total several times that today.[13]

During the first half of the twentieth century, few individuals in the American West considered the potential relationship between surface and groundwater. These two natural regimes were assumed to be independent, different beasts to be mastered. Surface flows could be seen and measured while groundwater could not. The latter's existence and movements were near magic and proven only by a producing well. Wildly inaccurate belief systems were conjured, describing underground lakes and rivers, which do exist in nature but in far different terrain and geological conditions. The occurrence of groundwater and its movement through interconnected pores among grains of sand and gravel was discounted, if even considered. Myth preceded a science not yet born.

During the latter part of the nineteenth and early twentieth centuries, wells for municipal use were completed for the valley's larger communities—Alamosa, Monte Vista, and Del Norte—when the towns accepted the need to provide a reliable water supply for their homes and businesses. Individual homes in the valley mostly had private wells. The question thus arises about the amount of water used by humans through individual homes, gardens, and businesses, plus the much larger consumption generally associated with municipal wells, when compared to total agricultural water consumption.

The answer to this question takes two forms. First, the total amount of water consumed by hundreds or thousands of humans and their settlements is miniscule in comparison to the amount diverted and consumed by hundreds of thousands of acres of agriculture. Contemporary water use in the United States approximates one hundred fifty gallons per person per day for all indoor uses, including drinking, meals and cooking, and bathing, plus car washing and garden and lawn watering. In the nineteenth century, San Luis Valley per capita consumption was probably far less. Ten thousand valley residents at the end of the nineteenth century might hypothetically have used 547,500,000 gallons of water per year, a seemingly huge number that is actually less than 1680 AF, or about as much water as the Rio Grande Canal diverts at full flow in *twelve hours.*

A second way of comparing human consumption to irrigated agriculture is to consider that roughly 10 percent of human use is permanently lost to the hydrological system in the San Luis Valley, or, if you will, the valley's natural plumbing. Stated differently, 10 percent of the water diverted for human use is actually consumed by the human body for physiological and related needs. The remainder—toilet flushing, bathing, and wash water—flow through an urban or rural treatment environment and returns to the valley's hydrological system by septic tank leach fields, municipal wastewater treatment plant discharges, lawn and garden runoff to streets, and municipal storm drains that discharge to nearby streams and rivers. By contrast, perhaps 30 percent of irrigation water is consumed by evapotranspiration and removed from the hydrological system. If more efficient irrigation practices are used, such as center pivot sprinklers, 70 to 80 percent or more of the water applied to the crops is consumed by evapotranspiration. The balance returns to the hydrologic system through surface and subsurface flow, or what the irrigation community calls *return flow.*[14] Return flow to another ditch, canal, or stream across the surface can take several hours to several days. However, if return flow is through the subsurface by means of the groundwater system, it may take several days to as much as several months or more before subsurface return flow returns to a receiving stream or river, or a groundwater well pumped for irrigation uses. The timing of a return flow's availability for re-diversion and application to crops is critical, because subsequent users may depend on return flows for their individual water rights. Water is diverted, applied, then subsequently made available to other appropriators by return flow, and the process repeats itself. The same water can easily wind up being "used" and "reused" (excepting transpiration losses) several times between Del Norte and the New Mexico state line.

Surface water from the Rio Grande and the valley's tributaries were the primary source of irrigation water through the first decades of the twentieth century, but eventual recognition that a groundwater resource was also available meant that the future would only prove more complex and interesting. Both the unconfined and confined aquifers gradually became increasingly important to the San Luis Valley in the twentieth century.

CHAPTER EIGHT

THE COMPACT AND THE CLOSED BASIN

In the Upper Rio Grande Basin, the use of water for irrigation constitutes practically its entire use.

— National Resources Committee (1938)

T he 1896 embargo, the Rio Grande Convention of 1906, and construction of the Elephant Butte Dam in New Mexico Territory were all driven by the desires of non-Colorado powers to control and apportion the Rio Grande's flows. As early as the 1890s, there was sharp disagreement among Colorado, New Mexico Territory, and Texas over how much Rio Grande water Colorado was fairly entitled to. The embargo temporarily mollified New Mexico Territory and Texas, but the problem was not solved because San Luis Valley farmers were still able to construct eight private reservoirs before 1920. Finally, in 1923, the federal government decided that a permanent solution to the Rio Grande apportionment ruckus would benefit all parties, and it recommended that the states and the federal government negotiate an interstate compact to formally apportion the river. It was hoped that a compact would minimize the effects of periodic water shortages that plagued the watershed and, if nothing else, the states might find a way to equitably share the intermittent pain.

The federal government was also looking for a reliable source for the 60,000 AF promised to Mexico in the 1906 agreement. During wet periods, the Rio Grande provided enough to satisfy most demands on the river, but as populations grew in the three states and new farms and ranches rose from the dirt, the combined demands of all three states and Mexico could not be met during droughts or dry late summer months. Complicating matters was the lack of reliable data on the river and its major tributaries.

All parties agreed to begin negotiations in 1929. There was little trust among the states, so a temporary compact was agreed to with the expectation that it would remain in effect until 1935, or until a final compact was signed. The temporary compact stipulated that Colorado and New Mexico would not increase their diversions from the Rio Grande during negotiations. It was, in effect, a standstill agreement.[1] For its part, and to assist the effort, the federal government assembled an interagency collection of technical experts, dubbed the National Resources Committee, and put it to work collecting data and evaluating the Rio Grande, its known and potential water yield, and agricultural practices, opportunities, and limitations.

Even with considerable input from the federal experts, final agreement on the Rio Grande Compact was not reached by the 1935 deadline. Disturbed by how long the process was taking, President Franklin Roosevelt issued an executive order prohibiting the federal government from approving any applications for new projects involving federal lands and waters of the Rio Grande.[2] Whether or not this helped, negotiations continued.

The National Resources Committee completed its studies and submitted its report in late 1937. The Rio Grande Joint Investigation was the most comprehensive and up-to-date collection of information on the Rio Grande available. One of its significant conclusions added to previous concerns about the San Luis Valley's use of *bringing up the sub* and elucidated several advantages and disadvantages. Disadvantages included over-diversion of streams and the Rio Grande during spring runoff, unduly high water tables, and excessive evaporation and transpiration losses. A specific advantage was that subirrigation substituted underground storage for "headwater," or storage higher in the watershed, thus better balancing water supply with irrigation demands.[3] This was important. It was the first formal recognition that the valley's irrigation future might depend less on reservoirs in the mountains and more on the valley's shallow aquifer. A solution to the valley's storage needs might be beneath farmers' feet.

Similar to what W. W. Follett and his report to the International Boundary Commission had concluded in 1896,[4] the Rio Grande Joint Investigation concluded that practically the entire normal flow of the Rio Grande and other streams entering the San Luis Valley from the San Juan Mountains was being diverted for irrigation,[5] affirming what New Mexico and Texas had been complaining about for more than forty years. Fairly apportioning the Rio Grande would challenge the status quo. Colorado had much to lose.

The Rio Grande Compact was finally signed in March 1938. Formal ratification required the states and Congress to individually approve the agreement, thereby making the compact law at the federal level and in the three states. This was completed by May of the following year. Whereas the 1929 temporary compact was a standstill agreement among the states that benefitted Colorado, the final Rio Grande Compact of 1939 allocated the Rio Grande in a manner less to Colorado's advantage. Given Colorado's cavalier practices for more than fifty years, this likely disappointed only Coloradans.

The Rio Grande Compact remains the single most important document guiding how the river is shared among Colorado, New Mexico, and Texas. For Colorado, the compact provides clear directives on how much of the Rio Grande's annual headwaters flow must be allowed to pass through to downriver states. In years when the Rio Grande flows abundantly, Colorado may keep more for itself; in dry years Colorado keeps less. Depending on annual snowpack and runoff, between 20 and 60 percent of the water generated in the upper Rio Grande watershed is allocated to the downstream states.[6]

Colorado and the San Luis Valley finally had confidence in how much water they controlled. Because water rights had long been the legal province of the states, the federal government had little influence over intrastate water rights, the exception being those obtained for specific federal projects or land, such as national parks and forests. Federal agencies have to comply with state procedures to attain such water rights. All this was extremely important to San Luis Valley farmers because, as concluded in the 1937 Rio Grande Joint Investigation, irrigation constituted nearly the entire use of the river in the upper watershed.[7]

One might wish to believe that with the Rio Grande Compact finally in place, the three states would behave honorably and adhere to the compact's apportionment of the river. But this is the American West, where water arguably is the most valuable natural resource. Agriculture, commerce, and government all covet water. In 1952, Colorado under-delivered its compact obligation by 153,300 AF, a sad pattern that continued through the major drought years of 1953–1956 and into the 1960s. By 1966, Colorado's accrued debit reached 944,400 AF, a whopping amount of water by anyone's standard. After Texas and New Mexico threatened litigation before the U.S. Supreme Court, Colorado relented, began to comply with its compact delivery obligations, and slowly started whittling away at its water debt. Fortunately for Colorado, the Rio Grande's runoff in 1985 was well above average and forced the first spill in over forty years from the Elephant Butte Reservoir in New Mexico. Under the compact's terms, Colorado's accrued debit, still in excess of 500,000 AF, was eliminated. Colorado has been in continuous compliance with its compact obligations ever since, part of which has been achieved by curtailing junior water rights (i.e., closing ditch headgates) when necessary to meet Colorado's delivery obligations.[8] Today, San Luis Valley irrigators half-jokingly claim the division engineer delivers Rio Grande water to New Mexico so precisely that he uses an eyedropper.

The 1929 temporary compact also began the process to provide 60,000 AF of Rio Grande water from the San Luis Valley's Closed Basin that had been promised to Mexico. The Closed Basin covers more than twenty-nine hundred square miles. Early explorers all observed that open water and soggy lowlands made crossing it with wagons nearly impossible. Crossing it on foot or by horseback was not much easier. The impossible terrain was due to the basin's internal drainage, recognized as early as the 1860s by Ferdinand Hayden.[9] C. E. Siebenthal, a geologist who surveyed the valley's water resources years later, concluded that the Closed Basin owed its existence to a hydraulic divide created by incipient sand dunes lying between the toe of the Rio Grande's alluvial fan and the valley's eastern slopes. The dunes created the Closed Basin by separating San Luis Lake and its neighboring streams and playas from the Rio Grande.[10] At some point in geologic history, Siebenthal and others believed that San Luis Creek flowed to the Rio Grande. However, the valley's strong seasonal winds had, over time, deposited sand dunes in this low spot on the valley floor and created the hydraulic divide and the Closed Basin.

Ideas about how to collect water from the Closed Basin and transport it elsewhere predated the 1929 temporary compact. All started with the premise that the basin's water could be put to best use by returning it to the Rio Grande. Accordingly, an investigation into the practicality of salvaging water from the Closed Basin was incorporated into the compact negotiations, an inviting concept because it would contribute to the conservation of water in the upper Rio Grande. In the 1930s American West, conservation notably meant something different from what it does today. Then, conservation almost universally meant "wise use," with the emphasis on use. The thought that humans should not use water to the maximum extent possible, that some water might be left in the river or for other economic uses, crossed very few minds.

The 1937 Joint Investigation concluded that the Rio Grande watershed's normal yield was fully appropriated and that only three sources of "new" water might be available: imported water from transbasin diversions; flood flows spilling from Elephant Butte Dam; and "new" water drained from the Closed Basin.[11] Because of the basin's interior drainage—water flowed in but seemingly went nowhere—water from the Closed Basin was declared to be "new." Runoff that naturally fed Saguache and San Luis Creeks in the spring supposedly flowed to the Closed Basin's low areas and then either evaporated or was evapotranspired by native *phreatophytes*.[12] Yet the Closed Basin also raised uncertainties. Plenty of water entered the basin, but if there were no surface outlets, where did it go? Could it all be lost to evapotranspiration and evaporation from San Luis Lake and surrounding playas? Evapotranspiration was the bugaboo of the era and remains so today. It was long blamed for unmeasured water losses from the Closed Basin, just as it has been blamed for water shortages throughout the American West. If water cannot be accounted for, blame evapotranspiration. Subsequent field investigations in the basin demonstrated that it is not so clear-cut. Evapotranspiration uses by native phreatophytes cannot account for all runoff that collects annually in the Closed Basin's low points, said low points being in or around San Luis Lake and locally called the "sump."

The National Resources Committee concluded that two sources of water supply the Closed Basin. The larger of the two was streams draining the Sangre de Cristo Range on the east side of the San Luis Valley. The smaller contribution was "from the west by the ditches and drains carrying waste and return flow from the area irrigated by diversion from the Rio Grande."[13] An engineered drain for the Closed Basin would satisfy Mexico's water demands by reclaiming water that irrigation in the valley did not consumptively use. This inference revealed a hydrological sleight of hand. The Closed Basin's salvaged water would be, at least in part, return flow from the valley's irrigation system that originated with the Rio Grande, and therefore was not really "new" water. But if Closed Basin water could be harvested and returned to the Rio Grande, nearly three thousand square miles of the valley floor would benefit by allowing irrigators to divert and use an equal amount of Rio Grande flow otherwise mandated to satisfy Colorado's compact obligation. It had all the makings of a win-win scenario.

The final Rio Grande Compact allowed Colorado to salvage "new" water from the Closed Basin, turn it back to the river, and credit it toward the state's obligation to downriver states and Mexico. Project proponents calculated that a Closed Basin Drain might annually salvage as much as 300,000 AF and deliver it to the Rio Grande. It was optimism on steroids. Later studies concluded that much less water might be salvaged, perhaps as little as 40,000 AF per year.[14]

Engineering studies for the Closed Basin Drain began shortly after the Rio Grande Compact was ratified in 1939, but they were shelved at the onset of World War II and remained dormant for thirty years. Congress got around to authorizing a Closed Basin project in 1972; however, construction did not begin for another eight years. As finally built, the drain roughly parallels San Luis Creek to San Luis Lake, continues southward to cross U.S Highway 160, then runs through the Alamosa National Wildlife Refuge to the Rio Grande. During the forty-plus years between the Closed Basin Drain's first formulation and its construction, the project mutated from a simple drainage canal to a forty-two-mile conveyance channel, known today as the Franklin Eddy Canal,[15] linked to 115 miles of pipeline laterals, fed by 110 pumping wells, and monitored by 82 observation wells. The drain's purpose morphed as well. In addition to providing "new" water to the Rio Grande to help meet compact commitments, the Closed Basin Drain supports the Alamosa National Wildlife Refuge and Blanca Wildlife Habitat Area and helps stabilize water levels in San Luis Lake.[16] The Closed Basin Drain finally began delivering water in 1988, although it was not fully operational for almost another ten years.[17]

To oversee and coordinate its obligations to the Rio Grande Compact, the State of Colorado created the Rio Grande Water Conservation District in 1967. In anticipation of a bonanza of salvaged water from the drain, the district initially applied for a conditional water right of 117,000 AF per year, admittedly far less than the initial projection of 300,000 AF. Reality never being as clear-cut as it was first drawn up, engineering difficulties with the drain and its components interfered and the district voluntarily reduced its annual water right claim to 83,000 AF, of which only 43,000 AF have been decreed absolute by a Colorado water court. The remaining 40,000 AF remain a conditional right and must be proven to exist before the court can declare it absolute. As of 2000, the Closed Basin Drain averaged an annual delivery to the Rio Grande and federal wildlife refuges of 17,300 AF per year.[18] During a severe recent drought, it produced about 14,000 AF.[19] Mother Nature is such a prankster.

CHAPTER NINE

COMPETING INTERESTS BUTT HEADS

*The water users of the Valley, often acrimonious in their
dealings with one another, came together almost as one in
their opposition to the AWDI application. They perceived the
application as a tremendous threat to their well-being and
continued existence.*

— Robert W. Ogburn (1996)

The meeting room filled slowly, noisily. Forty friends and colleagues, women and men, greeted one another with earnest handshakes and sociable nods. The first arrivals commandeered chairs and sat elbow to elbow at folding tables. Latecomers leaned against the walls. Everyone shed jackets and sweaters. The room smelled of wool, Carhartt, and wet earth. Spring had arrived in the San Luis Valley. So had the wind. Valley residents had grown accustomed to breakfasts of fried eggs and oatmeal seasoned with grit.

To no one's apparent dismay the hubbub continued past the appointed meeting time until, twenty minutes late, Mike Gibson, a compact man with close-cropped gray hair, stepped to the front of the room and called to order the monthly meeting of the Rio Grande Basin Roundtable. His British accent was apparent, but everyone had heard him speak many times and his diction was inconsequential. Gibson described the roundtable's composition. State and federal agency representatives were reminded that their roles were advisory only. Visitors were asked to identify themselves and their affiliations.

Formalities finished, an envoy of the Division 3 Engineer reported that spring runoff was predicted to be 51 percent of average, a statistic that elicited murmurs and groans around the room.[1] The worst drought in recent history was entering its second decade. Red dust had been observed on snowfields in the neighboring San Juan Mountains. Carried by prevailing winds from neighboring Utah, the dust, darker than snow, absorbed sunlight, warmed the snow, and stimulated snowmelt, leading to an early spring runoff. More groans crept through the room. Everyone wanted the mountain snowpack to melt slowly, take its time filling the watershed's reservoirs. Too-rapid runoff would disappear into New Mexico before Colorado could capture and store its share.

Steve Vandiver, general manager of the Rio Grande Water Conservation District, spoke next. A heavyweight in all matters related to the valley's water, Vandiver is well informed and respected. He reported that a computer model of the San Luis Valley groundwater system was still being fine-tuned in Denver and was expected to be working before the end of the year. The model was important; it would help the division engineer manage competing surface and groundwater rights. Irritation and confusion rippled through the room. Why was it taking so long? It was supposed to have been working months ago. How about an explanation? Vandiver shrugged his sizeable shoulders. The model was a first of its kind for the valley, and to explain it simply and succinctly seemed beyond his working knowledge of computer models. No one else in the room knew more than Vandiver. The model was overdue and needed. The lack of progress frustrated everyone.

More flotsam floated up, including the scary proposition from a government researcher that the current mountain snowpack, only 60 percent of the historical average, might be the new normal. More groans and muttering. Oil and gas drilling in Rio Grande County were mentioned. A proponent of subsurface drip irrigation requested funding for a demonstration project in potato fields in the valley north of Center. Travis Smith stood to describe the condition of the reservoirs in the upper Rio Grande watershed. They were aging. Whether publicly or privately owned, most were prohibited by state dam safety engineers from storing their designed capacity until badly needed repairs and rehabilitation were performed. Where would the money come from to renovate these dams and reservoirs? Silence. Smith continued. Governor Hickenlooper wanted statewide roundtable participation in furthering preparation of the state water plan. Each of the state's nine roundtables was expected to identify specific projects that could bring its basin's predicted water supply into alignment with predicted needs. For Colorado's Rio Grande watershed, completing dam and reservoir rehabilitation and moving forward with groundwater subdistricts were identified as critical to ensure a viable water future for the valley.

Next on the agenda was a status update on the state water plan. Greg Johnson, a water planner with the Colorado Water Conservation Board (CWCB), the state agency tasked to lead Colorado water policy and directed by Governor Hickenlooper to oversee the plan's creation, addressed the room. The ongoing statewide drought had stimulated public interest in the planning process. Colorado's agriculture community was especially worried about the drought and was questioning how the plan might help or hinder the state's farms and ranches. Johnson then unleashed a surprise. Governor Hickenlooper was replacing Jennifer Gimbel, the popular director of CWCB. Most in the room expressed surprise, some muttered disapproval. Vandiver in particular seemed unhappy, yet everyone knew that the head of CWCB served at the governor's pleasure. The roundtable body swiftly agreed to send a letter to Gimbel thanking her for her efforts on behalf of Colorado's water users. There was nothing more they could do.

Heather Dutton, coordinator for the Rio Grande Headwaters Restoration Project, spoke next. A tall young woman with a ready smile and quick intelligence, she is from generations of San Luis Valley farmers. After collecting several degrees at Colorado universities, she had returned to the valley to continue the family heritage; not on the family farm, though, but by working to improve the river's aging and inefficient irrigation diversions and to combat unnecessary erosion. She commenced her update with a frank declaration: the Rio Grande was struggling to transport more sediment than there was water to move it. The crux of the problem was the lack of water, not a surfeit of sediment, because the river was being largely diverted for irrigation. Sediment was finding its way into diversions and was causing problems in irrigation canals and ditches,[2] but the larger issue was the balance between river flow and sediment. If the two were in rough equilibrium, the river would be able to pass the natural and unending supply of sediment downriver rather than allow it to aggrade the river channel. All rivers perform this function, some better than others. But the sediment load in the Rio Grande had the upper hand. The valley's major irrigation canal headgates were plugging with sediment. On the positive side, Dutton reported that the floodplain, the link connecting the river's riparian corridor with the river itself, was gradually improving. River water quality was also getting better.

The organization Dutton headed had been in existence barely a dozen years. It represented an amalgamation of interests and input from groups and individuals that had come together to address the river's decline as a self-sustaining hydrological system. The organization began in 2001 by commissioning a technical report that addressed conditions on ninety-one miles of the river, extending from South Fork downriver to the Rio Grande's confluence with La Jara Creek. Over this reach, the river is affected mostly by irrigation diversions, ranching, and wildlife habitat. Recreational use of the river is minor because of limited public access.[3] The report provided an overview of the river and its uses over the past hundred and fifty years, and documented how the river and its associated riparian corridor had changed as a direct result of agricultural practices and irrigation withdrawals.

Like all rivers, the Rio Grande autonomously adjusts its channel to accommodate the amount of water it moves. If climate changes, flows change in response and channel sizes adjust. The process is gradual and natural. Most river flows, though not always in the West, increase in the downriver direction as tributaries and groundwater inflows add water. But subtract significant amounts of water for irrigation and over time a river channel will decrease in size through sediment deposition, called *aggradation*, either in width or depth or both. The Rio Grande's bank-full capacity near the New Mexico state line is less than 500 CFS, while upriver it may be as much as 8,000 CFS.[4] The difference is due almost entirely to irrigation diversions. Natural, undammed rivers flood, on average, every other year. Flooding along the Rio Grande had been a nonissue during the ongoing drought, but if flooding were to occur between South Fork and Monte Vista it would threaten irrigation diversions along the river and on the floodplain.[5] Municipal and suburban development on the Rio Grande floodplain had put infrastructure, including bridges and levees, not to mention public safety, at risk.[6] The same technical report concluded that "None of the land use planning documents and ordinances currently being used…in the Rio Grande corridor provide adequate ability to protect and preserve the corridor from adverse impacts of future…development."[7] The Rio Grande is not the only river where such risks exist. Human developments on major floodplains all over the world have increased the risk of significant flood damage. Decisions allowing such development are as common to this nation as anywhere else.

Native vegetation on the Rio Grande floodplain has also been sharply affected by irrigation diversions over the past hundred and fifty years. Above Monte Vista, the floodplain and riparian corridor are dominated by mature cottonwood trees. New growth has been limited, however, by grazing and the lack of periodic flooding that allows cottonwood seeds to germinate. Especially since the early 1940s, the decline of cottonwood galleries along the river has been attributed to agricultural and urban development. Once farm fields and subdivisions crowd a river, they are rarely removed in favor of restoring trees.[8]

Below Monte Vista the riparian cottonwood galleries have thinned. Dead trees stand as decaying specimens, likewise due in large part to infrequent flooding. By the time the river reaches the Alamosa National Wildlife Refuge, flourishing cottonwoods are rare and the river enjoys little shade from mature trees. Those that remain are leafless, jagged skeletons, long dead. Willow thickets line the river instead of taller trees. It has been a long time since a major flood wetted this reach of floodplain.

Throughout the two-hour meeting, discussion was civil and peppered with humor and personal asides. Everyone in the room had a personal and vested interest in the river, the valley, and the communities they supported. Nonetheless, roundtable members did not speak with a single voice or opinion. The manager of the San Luis Valley National Wildlife Refuge Complex, a nonvoting roundtable member, was addressed with deference not commonly accorded federal water rights owners elsewhere in the West. His responsibility was waterfowl, not potatoes or alfalfa. Rio de la Vista, one of the roundtable's leaders, was a consultant to the Rio Grande Headwaters Land Trust, an organization devoted to acquiring and improving wetlands in the valley. Perhaps more interested in marsh birds and cattails, she was rarely in complete agreement with many of the valley's agricultural water users, but she understood the valley's economy and its dependence upon the river. The roundtable was a blend of divergent voices.

Many farmers in the American West sincerely believe they have a moral right to the region's limited water, or at a minimum first dibs on its availability. This belief dates to the region's first settlers who arrived before there was anyone to challenge them. They came, they saw, they took, a modus operandi exercised throughout the West. Many Colorado farmers, as well as literally thousands of their regional brethren, zealously contend that water not used to grow crops or support livestock is literally wasted. It is not uncommon to hear an opinion that water flowing un-diverted in a river or stream is squandered.

To be fair, this belief system did not originate with western settlers. As far back as Aristotle, humans justified or believed that nature existed for the sake of humankind. Eighteenth- and nineteenth-century European philosophers continued in the Greek philosopher's footsteps.

North Americans followed suit. Nature was intended to be subdued and developed; it was our duty.[9] Turning the resources of the untrammeled American West to the wishes of man was simply a logical extension of this thinking. Little has changed.

By contrast, many environmentalists, serious ecologists, and fans of outdoor recreation are dismayed by such conviction. They want to see clean water flowing in streams and rivers and supporting healthy riparian corridors, abundant fisheries, and greater recreational opportunities. Regardless of these differing ideas about how to use water, however, one certainty is that, under Colorado law, agriculture holds the legal rights to most of the water and its use. Farmers arrived first and put the water to beneficial use. Their right to it is nigh impossible to challenge. If anyone at the roundtable meeting held a differing opinion, they failed to voice it. An outsider would have been hard pressed to distinguish personal biases based on behavior and language. Everyone at least pretended that his or her interests and goals were compatible. Not that they agreed on everything, of course, but their respect for and reliance upon the Rio Grande had brought them together to solve common problems. The San Luis Valley has benefitted from a hard-fought battle that brought its residents together.

◆ ◆ ◆

In 1986, American Water Development, Inc. (AWDI), a band of investors and speculators, surprised the San Luis Valley's water community with a proposal to annually export up to 200,000 AF of groundwater to thirsty Front Range cities that lay well beyond the physical and hydrological limits of the valley. For the first time, a scheme was proposed to mine and market water on a scale vastly greater than anyone in the valley had ever considered.

The conflict began with AWDI's application to the Colorado Division 3 Engineer for a conditional groundwater right. The application identified where wells would be constructed, how much water would be pumped, and how and where the pumped water would be beneficially used. Beneficial use is defined by state law. Municipal use, even halfway across the state, meets the standard.

Following receipt of an application, professionals in the engineer's office determine whether decreed water rights may be affected. If that hurdle is cleared and no objections are filed—a rare instance in itself—a water court judge customarily approves a conditional right. Only at that point may applicants legally complete the well or wells and pertinent infrastructure to put the water to the proposed beneficial use. After the wells are completed, the pumps and pipes installed, and water delivered to an end-user that meet the criteria for a beneficial use, AWDI would be in position to request a decreed water right from the court. Under Colorado law, a decreed water right possesses the certainty of an established property right.

To no one's surprise, AWDI's application brought out numerous objectors, including the State of Colorado and several federal agencies, all claiming that AWDI's proposed 112 wells near Villa Grove in the valley's north end would materially harm their decreed water rights, that the wells "would suck the lifeblood of the Valley out of the ground and turn their verdant valley into a...wasteland."[10] The Division 3 Engineer eventually supported the objectors' claims that their water rights would be impacted and declined to recommend approval to the water court. Denials can result in project proponents rolling up their hoses and going home, or they may trigger negotiations among the parties and modifications to the project proposal until all parties are satisfied that their interests are protected, at which time the application is approved and forwarded to the water court. But AWDI's proposal was far from a typical application. Valley objectors opined vehemently that AWDI's plan was nothing less than a declaration of war on their way of life, to say nothing about threats to their water rights. Negotiations would solve nothing. If AWDI wished to proceed, it had no recourse but to file suit in Water Division 3 Court. Let the judge decide.

The legal challenge before AWDI was to demonstrate to the judge's satisfaction that the Division 3 Engineer was wrong, that AWDI's project threatened no one, and that senior decreed water rights would be unaffected. Experts were hired. Additional field studies and computer modeling were completed and submitted to the opposing parties and the court. Experts were deposed to explore the other side's case and expose its technical strengths and weaknesses, and to demonstrate to everyone—if any doubt previously existed—that lawyers knowledgeable about Colorado water law can make a healthy living arguing cases in court. The western axiom that "Whiskey is for drinking, water is for fighting" was again proven true.

The gist of AWDI's argument was based on its interpretation of the valley's groundwater system and how much groundwater could be made available without impacting extant rights. AWDI's experts grounded their case on the belief, deemed by many at the time to be reasonable, that the valley contained two separate aquifers: the manmade, near-surface, unconfined aquifer that had been created by irrigation over the preceding century, and an underlying confined aquifer that was the source for the numerous flowing wells and springs around the valley's perimeter. Geological studies going back to Ferdinand Hayden's surveys in the 1870s described a vertical sequence of widespread fine-grained lakebeds, locally called "blue clays," that was believed to separate the two aquifers. And where the upper, unconfined aquifer was annually refilled (*recharged*, in hydro-speak) by the valley's extensive network of leaking irrigation canals and ditches, AWDI maintained that the deeper confined aquifer was recharged by the numerous mountain streams that emptied into the valley and were thus independent of the river and most valley irrigation. AWDI calculated that its wells could handily pump 200,000 AF per year from the confined aquifer without significantly affecting the overlying unconfined aquifer, the valley floor's network of irrigation canals and ditches, or the Rio Grande. The two aquifers were, in AWDI's opinion, independent and separate.

AWDI's case also relied on an estimate of the total volume of groundwater contained in the confined aquifer, a position that relied in part upon work undertaken by one of the early investigators of the valley's hydrologic system, Philip Emery, of the U.S. Geological Survey. Emery admittedly and simplistically estimated that the confined aquifer might contain as much as 2 *billion* AF of groundwater, or approximately ten thousand times the amount of water AWDI wanted to export annually.[11] A separate and more optimistic estimate put the volume of groundwater in the confined aquifer at 65 *billion* AF.[12] Surely, AWDI argued, the amount of water it wished to export was insignificant considering how much water the aquifer contained. There would be no impacts on existing users and their decreed water rights, AWDI argued, adding that the confined aquifer was not tributary to the Rio Grande and would not impinge on Colorado's obligations to New Mexico and Texas under the Rio Grande Compact.

AWDI's opponents hired their own experts and attorneys. Although several counties in the San Luis Valley are among Colorado's poorest, a special election voted twenty-to-one in favor of a new property tax to help pay for litigation costs.[13] A majority of AWDI's opponents based their objections on the threat to agriculture and their rural lifestyle. Multiple irrigation companies, canal associations, individual farmers, and assorted plaintiffs of every stripe—most agricultural—banded together to battle AWDI. Backed by Colorado's prior-appropriation water law, they had spent more than a century solidifying their interests through the courts and state legislature. Agriculture required irriga-tion in the San Luis Valley. Rainfall would never satisfy the needs of most crops. Farmers feared their livelihood was doomed if AWDI won in court. The valley economy depended on irrigation. Federal and state agencies also felt threatened by AWDI's proposal and joined with the valley's agricultural water users to fight the water export proposal.

With so much at stake, the legal process took considerable time for the experts and lawyers to prepare. Their preparations included review-ing the opinions of experienced consulting engineers and hydrologists, as well as studying their calculations and finely tuned groundwater computer models. Five years later, following a six-week trial, the Water Division 3 Court ruled in favor of the objectors and dismissed AWDI's claims and application. AWDI subsequently lost an appeal to the Col-orado Supreme Court and lost again when the U.S. Supreme Court refused to even consider an appeal. A year later, Colorado's U.S. Sen-ator Tim Wirth maneuvered a clause onto a Congressional water bill that barred any groundwater development in the San Luis Valley that would damage the Great Sand Dunes National Monument or other fed-eral projects, like the U.S. Bureau of Reclamation's Closed Basin Proj-ect or wildlife refuges managed by the U.S. Fish and Wildlife Service.[14] Objectors eventually recovered over three million dollars from AWDI to satisfy trial court judgments to compensate them for their battle's expenses.[15] AWDI later sold its interests in the San Luis Valley to other speculators and developers who still retain an interest in developing the valley's groundwater. Few valley residents argue that all water export schemes are dead. Where money might be made by developing or mov-ing water, the threat is constant.

The public relations and court battles surrounding AWDI and its proposal have been ably recounted in Sam Bingham's *The Last Ranch*,[16] an eloquent description of farm and ranch life in the San Luis Valley at the end of the twentieth century. Bingham's tome helped cauterize the valley's bickering water interests and keep them together. Those with the most to lose were farmers who relied on irrigation and believed that pumping the valley's confined aquifer would sooner or later impact their irrigation wells and the valley's irrigation canals and ditches. Envi-ronmental interests joined with agriculture to fight the proposal. That the valley's ponds, sloughs, and wetlands would never be returned to what they were two centuries ago was not the issue. They simply did not want to lose what they have. Explanations of the valley's complex irrigation plumbing are far from simple, but the uncomplicated model that everyone understands is what transpires when a straw begins slurping at the bottom of a soda cup. The San Luis Valley is a mighty large cup, but its water resources are finite. There is a bottom.

During the five years that AWDI's water export scheme hung over the valley like the Sword of Damocles, the valley's various interests and their representatives learned that they had more in common than previously believed. If the episode had not quite created valley-wide affability, the AWDI episode at least forced recognition that cooperation among competing interests could work in their favor. When the Rio Grande Basin Roundtable came into existence in 2005, the valley's water interests found that they already had a structure capable of furthering their hard-earned communal sense of purpose.

CHAPTER TEN

A Fire and a Kitchen Sink

*I went to the water users' groups. I went to the canal
companies. I went to the San Luis Valley wetlands focus
group. I went to the Division of Wildlife. That started a
conversation about this whole idea of re-operating reservoirs,
multipurpose, multi-use.*

— Travis Smith, 2013

Travis Smith's Ford pickup is his desk on wheels, a four-wheel-drive office. During the irrigation season, from April into September or October, he daily drives the miles of dirt roads that lace together the sixty-two thousand acres comprising the San Luis Valley Irrigation District. His cell phone is equally important. It allows him to tackle the multitude of problems and answer the many questions that crop up daily while he simultaneously directs his ditch riders to open and close the headgates that control water delivery to the district's hundred and fifty members.

Smith will not be driving around the valley today. He has a meeting at the Rio Grande Dam, high in the San Juan Mountains, a two-hour drive from Center, and he hurries to be on the road. He folds his tall frame behind the wheel and stretches across the front seat to shovel papers, maps, and fast-food wrappers into the back to make room for a passenger. Passenger aboard, he begins the journey.

Twenty minutes later, Smith stops briefly to show off the Farmers Union Canal headgate, a prominent steel and concrete structure, the point where the district's irrigation water is diverted from the Rio Grande. Minutes later he catches US Highway 160 heading west from Del Norte. Sixteen miles farther on, at South Fork, so named for the confluence of the Rio Grande's South Fork with its main stem, he turns onto Colorado Highway 149 toward the historic mining town of Creede. The highway zigzags and doubles back on itself before breaking into a broad, lush valley stippled with vacation homes and somnolent cattle separated by barbed wire fences. After a hamburger and fries at the Fremont Store, the only roadside café in fifty miles, Smith continues up the narrowing valley. It is a warm September weekday and traffic is thin. Most tourists left the mountains before Labor Day to get their children back to school. The few trucks and recreational vehicles on the highway belong to serious fishermen and vacationers who waited until after Labor Day before venturing into Colorado's high country.

Below Spring Creek Pass, Smith turns onto a forest service road and begins the climb to the dam. After several miles of bumpy driving, a charred and blackened forest appears. Smith pulls the truck off the road, stops, and gestures across the canyon toward the torched timber. "There's eighteen square miles in that basin that's all burned, a high-hazard, high intensity burn. They've already had some small debris flows, a lot of ash flow. A lot of sediment has already come off these slopes."[1] His concern is palpable. An intense wildfire leaves little or no vegetation to hold soil in place. If, or more likely when, ash and sediment from these scorched slopes find their way to the Rio Grande, the consequences may prove devastating to the river's aquatic life and fishery, not to mention downriver irrigation.

Three months earlier, on the afternoon of June 5, 2013, lightning sparked the West Fork Complex forest fire. Three separate fires at first, all lightning-triggered, grew aggressively until they coalesced and the U.S. Forest Service dealt with them as a single blaze. Upwards of sixteen hundred professional wildland firefighters from around the Rocky Mountain West were quickly assigned to the conflagration. The fire moved rapidly into remote and expansive areas of timber previously killed by the pine bark and spruce beetles. The bug-killed trees were little more than standing kindling and the flames took advantage. Summer heat and high winds made firefighting both difficult and dangerous. Because much of the fire burned within designated wilderness, the interagency fire management bosses chose to avoid assaulting the fire itself, instead focusing on protecting at-risk vacation cabins and homes in the national forest and on the fire's perimeter. The towns of South Fork and Wagon Wheel Gap were evacuated. U.S. Highway 160 and Colorado Highway 149 were temporarily closed. Crews worked around the clock to keep the flames away from inhabited areas while the fire spread deeper into adjacent roadless and wilderness areas.

Smith described the fire's advance up-valley toward his dam and reservoir and pointed to a rocky buttress across the canyon. "This was their last back-burn. Everything worked pretty well, and then all of a sudden a three-thousand-foot plume built, and that hot air hit that colder air above, and it collapsed and the wind changed." This was predictable behavior for extremely hot wildland fires. As a fire intensifies, hot air and smoke climb into the sky, fed by air pulled along the ground toward the plume in its own windstorm. But when the plume rises high enough to encounter cold air, the plume collapses, and everything reverses. The wind suddenly blows from the plume's base back toward the fire's perimeter. Smith said, "That hot air hit that cold air and it collapsed and the wind changed and ran all the firefighters out of there. They were all down here on the road. These were veteran firefighters. You could tell they looked a little shocked." He paused. "It's all this standing dead timber. The fuel load was just so high."

The professionalism of the firefighters who had battled the vast blaze impressed him with their willingness and ability to keep the fire from reaching his dam and reservoir. Had the fire surrounded the reservoir, the fire-denuded slopes would, perhaps for years, shed sediment, ash, and burned timber into the impoundment, lessening its storage capacity and quite possibly plugging its outlet. The reservoir could have been rendered useless. In Colorado and the American West, little spooks farmers and ranchers more than a threat to their water. Wildfires near their mountain reservoirs always constitute a threat.

Over subsequent weeks, Colorado's late summer "monsoon" season finally showed up. The skies clouded up in the afternoons and the weather grew cooler and more humid. Afternoon rains hindered the fire's progress as exhausted crews watched gratefully from its perimeter. Six weeks after the fire began, the last incident report was filed by the interagency fire management team. The report described the fire as "smoldering." Mop-up operations continued for several more weeks, and the number of firefighters shrank to forty-three. In the end, the West Fork Complex fire torched in excess of one hundred and nine thousand acres, a hundred and seventy square miles of chiefly national forest.

◆ ◆ ◆

The Rio Grande Dam comes into view, though someone expecting a grand mass of concrete, like New Mexico's Elephant Butte Dam on the lower Rio Grande, would be disappointed. Approached from below, the earth- and rock-filled structure looks more like a chunk of sod that has been plopped brusquely between the canyon walls. Water gushed from its base where the outlet discharged water to the aspen-hugged river. Smith pulls his truck onto the dam crest and stops next to a large granite boulder into which a message had been chiseled: *Rio Grande Reservoir, 1912–2012, 100 Years.* The boulder was placed there during the irrigation district's public celebration of the dam's first century.

In spite of this granite milestone, the largest dam and reservoir in the upper Rio Grande watershed has, according to Smith, "long-standing dam safety issues because it's built on a landslide. We have seepage issues that we've identified. We have a hundred-year-old outlet structure that has operating limitations just by its design."[2] The dam, as necessary and important as it is to the district's members and the San Luis Valley, was suffering from advanced age and the shortcomings inherent with century-old design and construction.

Because of uncertainties underlying construction of a major new dam and reservoir in the American West in the twenty-first century, Smith and his irrigation district had chosen to rehabilitate the existing structure. "You can't go out and buy a reservoir and you can't go out and build a reservoir," he said. Among many other hurdles to overcome in order to construct a major new dam and reservoir, funding such a project would create a circular firing squad of proponents, each insisting the others pay first and most. This does not include the formidable tasks of identifying a site and years of environmental and engineering studies and permitting required before a shovel could be lifted or a rock moved.

On that day, engineers, technicians, and heavy equipment operators are locating local deposits and processing sixty thousand cubic yards of natural clay for the dam's upstream face to control troublesome seepage threatening the dam's stability. The rumbles and roars of heavy equipment reverberate between the canyon walls, as the machines strip the surface soil and rock from the dam's upper face and replace it with a compacted clay layer. When that is finished, engineers intend to drill a series of holes in the dam face and inject grout into the dam's core. Together, the clay curtain and injected grout should reduce the seepage and allow the irrigation district to use the reservoir's full capacity without worrisome handwringing by state dam inspectors.

This portion of the dam's rehabilitation will cost on the order of five million dollars, provided by a grant from the CWCB to Smith's irrigation district.[3] But controlling seepage through the dam is just the first phase of the dam's new life. The second phase will entail major repairs to its outlet structure, for which another fifteen million dollars has already been approved by the state legislature and set aside by CWCB. Smith is optimistic that the final engineering designs for phase two will be forthcoming within months, and with luck he hopes contractors can begin work the following year. Smith emphasizes that physically enlarging the reservoir was not the original intent of the rehabilitation plan, although he mentions a hoped-for phase three to upgrade and raise the dam's spillway crest, thereby adding approximately 10,000 AF of storage capacity to the reservoir without increasing the dam's size or raising the dam crest.[4] The dam's original constructors had inadvertently made the dam high enough that Smith's team could raise the spillway and increase reservoir capacity without fundamentally altering the dam's structure.

Later, as he turns his truck homeward, Smith seizes the opportunity to tell the backstory to the Rio Grande Dam's rehabilitation. In the West, when water infrastructure is under discussion, especially when millions of dollars are involved, time is an independent, always difficult-to-corral variable. The story behind the Rio Grande Dam rehabilitation project is no different. "It all starts with an idea," he said.

Smith was a state water commissioner on the Rio Grande for fourteen years before he became superintendent of the San Luis Valley Irrigation District. Western irrigators never seem to have enough water, so little time passed before he began to wonder whether the capacity of the Rio Grande Reservoir could be increased, perhaps doubled in size. As he canvassed the valley for opinions and support, he realized that few individuals and interest groups favored the reservoir's expansion. Moreover, the financial burden of such a project frightened members of Smith's irrigation district. Political and environmental impediments might also be impossible to overcome. Smith is not naïve and recognized a dead end.

Out of this reality check was born the Rio Grande Reservoir Rehabilitation and Enlargement Study, or what Smith refers to as the "kitchen sink" study about what could be accomplished if the reservoir were enlarged or rehabilitated. Who in the upper Rio Grande watershed would be interested in increased water storage? Irrigators obviously, but did other individuals or organizations see a need for more storage? Smith visited as many individuals and groups as would listen. After several years, the study and discussions informed his conclusion that interest did exist, after all, for a 10,000 AF increase in reservoir capacity.[5] But who would pay for the project?

A necessary catalyst showed up in 2005 when Russell George, director of Colorado's Department of Natural Resources, wondered aloud what state government could do to help meet Colorado's future water needs. Most of the state's population growth was taking place on the Front Range, largely centered on Denver and Colorado Springs, while most of Colorado's water is on the opposite side of the Continental Divide, what Coloradans call the West Slope. As well, the state's largest rivers are nearly all fully appropriated, primarily for agriculture, and most private irrigation companies and districts would struggle to finance dam rehabilitation projects costing tens of millions of dollars. Smith paraphrased Russ George: "We are past the time when irrigation outfits, who own most of the reservoirs and who own most of the water [rights] in Colorado, are able to take on the tens of millions of dollars that it takes for infrastructure upgrades. What can the State of Colorado do to kick start, to support, to come alongside, different water user groups, *if they're willing?*"[6]

If they're willing was a critical distinction made by George. No one was going to force farmers or private water communities to team up with the state or anyone else. But for organizations like the San Luis Valley Irrigation District, state financial assistance *might* be available, with one very significant stipulation: infrastructure rehabilitation must serve multiple purposes and multiple uses. Future state grants and loans would go preferentially to entities willing to enter into partnerships and agreements that were about more than irrigation. No one would force irrigation companies and districts to change their operating protocol or habits. *If they're willing* became the operative guidance. For many agricultural water users around Colorado and the West, this would not be an easy sell. The new paradigm would constitute a major shift in the operating philosophy of those organizations who owned their dams and reservoirs. If anything characterizes western attitudes toward water, it is that agriculture gets its way, and that financing from state and federal legislators and agencies happens frequently. But no more.

The state legislature eventually agreed to provide money to put some of these ideas into practice. New guidance from the legislature and cwcb encouraged Smith and his board of directors to engage in some serious mind-stretching. The Rio Grande Dam had been constructed more than a century earlier solely for irrigation. Over the intervening decades, it had also been put to use for flood control and storage for other uses, always with the district's approval. On any given day, Smith might authorize releases for as many as eight different transactions involving stored water. But the dam had aged and was unable to safely store its full capacity. TLC was needed. If the district wanted financial assistance from the state, Smith and his board of directors would have to rethink how they might operate the reservoir, how they might retime reservoir releases, how they could achieve multiple benefits with the same drop of water.

Smith and the district rose to the occasion. In conjunction with the Colorado Department of Parks and Wildlife (CPW), they came up with what would become known as the Rio Grande Cooperative Project. The irrigation district agreed to virtually combine its water in the Rio Grande Reservoir with water stored in CPW's Beaver Creek Reservoir. The latter is located miles from the Rio Grande Reservoir on a tributary to the South Fork of the Rio Grande. Both reservoirs are in the Rio Grande watershed, but the sources of their water differ.

"The idea of partnering up really started with my district," Smith said, "because we saw that we could not take on a ten or twenty or twenty-five-million-dollar project." Irrigation districts in Colorado possess taxing authority and can plausibly bond infrastructure projects, then pay off the bonds by increasing taxes on the district's landowners. District members could pay for the dam's rehabilitation out of their own pockets, although this alternative understandably was not embraced with jubilation. For one thing, all taxpayers in the district would be saddled with bond repayments, not just the farmers and ranchers benefitting from the dam's repairs. More to the point, why pay for something yourself when the possibility exists to obtain grants or low-interest loans from the state?

Another key factor was the widespread and growing realization that the valley's transition to sprinkler irrigation, begun in earnest in the 1970s, made the timing of irrigation water releases from the reservoir almost irrelevant. Reservoir water could be released practically anytime, as long as it eventually got to the canals and ditches and was allowed to sink into the earth to recharge the valley's unconfined aquifer, which is far more important and much larger than the reservoir. Having water beneath their feet when farmers need to irrigate make its use substantially easier. There is no time lag between need, delivery, and use.

The irrigation district and CPW eventually agreed that they could operate their two reservoirs jointly for multiple benefits, including enhancing instream flows and providing for fishery and irrigation needs. All they had to do was figure out a way to get more than a single benefit from the same acre-foot of water by retiming and coordinating releases from the two reservoirs. Just as important, Smith emphasized, "You've got to have reservoirs that work to be able to do that. That is where the money comes in."

The general conclusion of the two organizations was that their differing needs and sources, when approached simultaneously and cooperatively, allowed complementary solutions. CPW had historically kept its Beaver Creek Reservoir full to use for recreational fishing by the public, rather than tap it for fishery or agency needs lower on the river. The agency had five transmountain water rights and diversions that could bring additional water into the Rio Grande watershed from west of the Continental Divide, but because Beaver Creek Reservoir was always kept full for fishing, CPW had no place to store additional water and could make no use of the diversions. CPW also owned shares in different ditch companies in the valley. Smith said, "They had this water portfolio, but they weren't fully exercising it because with Beaver Reservoir being full and they're not having a storage agreement with us, they were foregoing a lot of their water rights."

This was the state of affairs when Smith met with CPW's Jeff Johnson and began discussions concerning what the agency could do if it had additional storage. CPW eventually entered into a temporary storage agreement with the irrigation district to store water in the Rio Grande Reservoir. CPW also agreed to begin using its transmountain diversions and storing that water in the Rio Grande Reservoir for eventual release to the river. "Getting them wet water in the river," Smith said, "is good for wildlife. That helps them meet their mission." Among water professionals, *wet water* is the term applied to a senior water right. It is how water managers refer to water that will actually show up in streams, rivers, and reservoirs, as opposed to hypothetical water dependent upon junior water rights that are rarely fulfilled, called *paper water*. In Colorado, wet water is the only kind worth spending money on or caring about. Junior water rights are the last in line, dependent on climate, weather, and the return flows of senior appropriators upstream, or pumping from the same aquifer.

According to Smith, the cooperative agreement could not have been negotiated fifteen years earlier. Two years were required to put together a draft of the storage agreement, during which cooperation, trust, and personal relationships were developed. "That's what the Rio Grande Cooperative Project is," Smith explains. "We meet multiple benefits with the same water. It's the idea of taking both of these reservoirs and operating them for the maximum benefit. They still meet their mission. They still meet their decreed needs. It is the model that I think the rest of the state is going to share."[7]

Smith was less optimistic about the future of irrigated agriculture in Colorado. As of 1990, the state had more than 1,900 reservoirs capable of storing 885 *million* AF.[8] The Rio Grande Cooperative Project may prove to be a successful model for statewide improvements to both irrigation and fisheries, but rehabilitating a significant number of aging irrigation structures will require financial assistance on a grand scale. Taxpayer support might be limited, and federal largesse is increasingly thorny to come by. But if money cannot be found to repair the state's aging dams, the consequences are equally grim. Smith wondered aloud, "Do you just take land from $2,500 an acre to $200 an acre and have weeds grow? How do you do that without affecting the school districts, the tax base for individual counties? It's a huge effort."

Would CWCB have entertained the district's request for millions of dollars in grants and loans to mend the Rio Grande Dam if the district had applied independently, *without* including CPW and multiple uses as integral parts of the project?

"No way," Smith said.

Halfway back to Center, Smith turns off the highway onto a dirt road leading to a simple bridge over the river. Bridge timbers rattle as the truck crosses it. He stops the truck on the far side and walks down to the water. The September afternoon is fading, and shadows are lengthening across the Rio Grande. Lunker trout surely loiter downstream of sofa-sized boulders lodged in the current. The moving water runs clear as cold crystal, endlessly flickering in the dying light. It is what draws us to rivers, to the magic.

Smith bends down and traces his fingers along the water's edge. His fingertips come up black with ash and bits of charcoal. The wildfire fought months prior had released ash and burned timber into the river's upper tributaries, and now the detritus is finding its way downriver. Had the fire extended above the Rio Grande Dam, the resulting eroded soil, ash, and burned timber could be ruinous to the farming and ranching community. But if protecting the dam and reservoir is critical to their agricultural utility, just resurrecting those structures may not be sufficient. The San Luis Valley has fourteen reservoirs with 349,000 AF of storage, primarily though not exclusively used for irrigation. Rehabilitating at least some of them may be necessary.[9] Not all of these other dams require expensive maintenance and upgrades, but many do. Where the funding will come from to pay for the needed work is not obvious. Federal and state sources will likely be the first target for project proponents. Agricultural interests in Colorado have historically been successful in getting the state legislature to help clear financial hurdles. Agriculture is one of America's most sacred cows, but far fewer individuals claim the profession today compared to a century ago, and they credibly fear that their political clout wanes. Farming and ranching remain risky financial endeavors. Even with modern farm equipment, they are physically demanding occupations. The financial returns rarely permit paying for major water projects, even if an agricultural community is willing. Smith and his allies argue that additional cooperative agreements in the future will allow the greatest possible benefits from the same water, benefits that admittedly may not always accrue completely to agriculture. If public financial support is sought, agriculture may need to adopt parallel environmental benefits to attain the indispensable political and financial support.

Smith stares into the distance, stands, wipes his palms together, and strides toward the truck. He has another hour to drive before home and dinner.

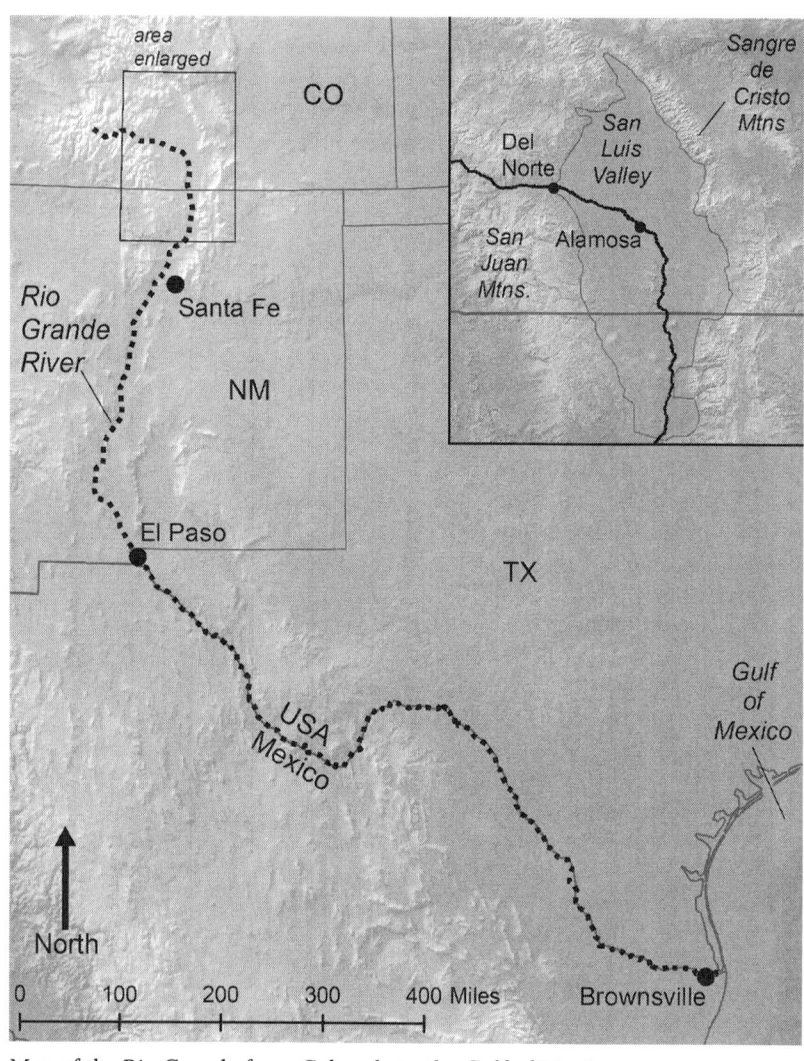

Map of the Rio Grande from Colorado to the Gulf of Mexico

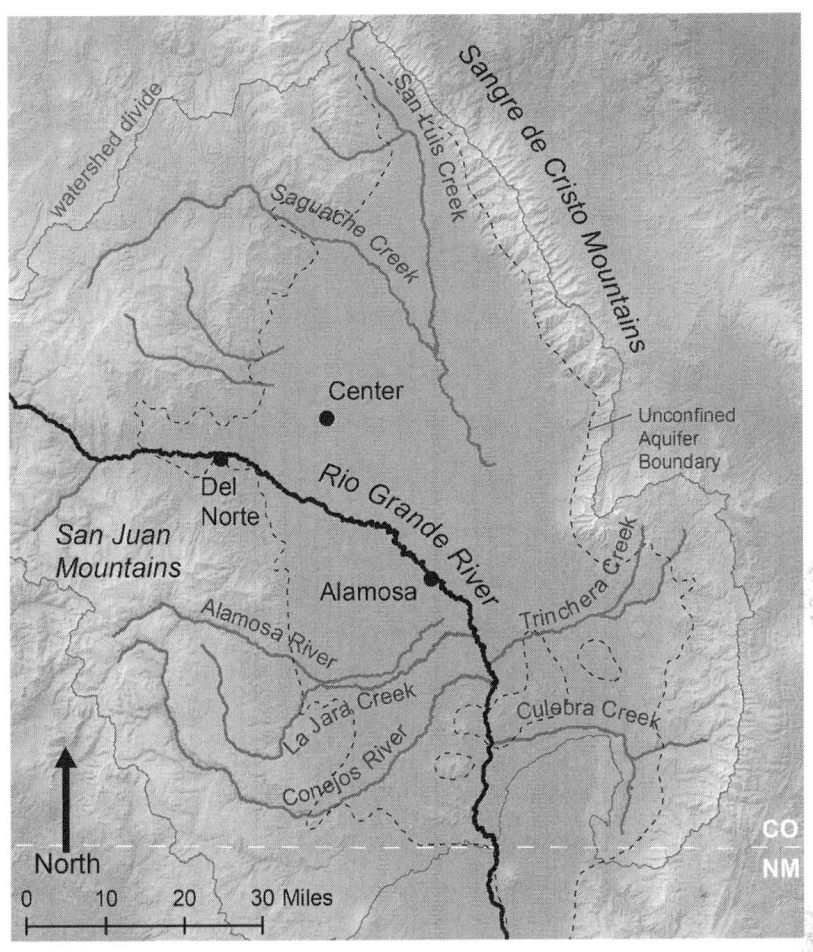

Map of San Luis Valley and Upper Rio Grande

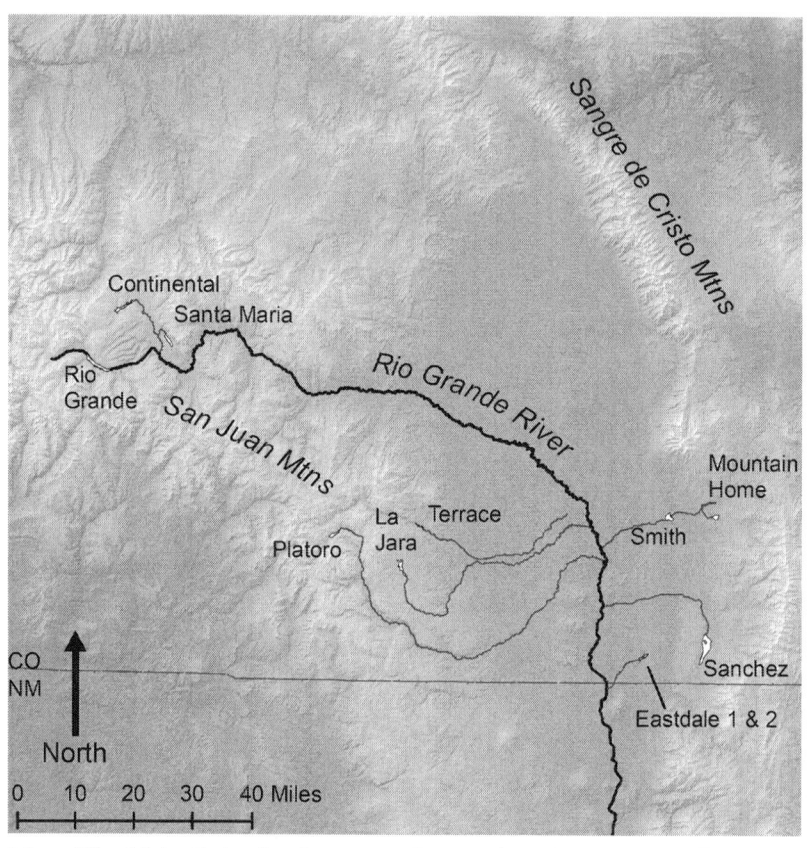

Map of Ten Major Irrigation Reservoirs Surrounding the San Luis Valley

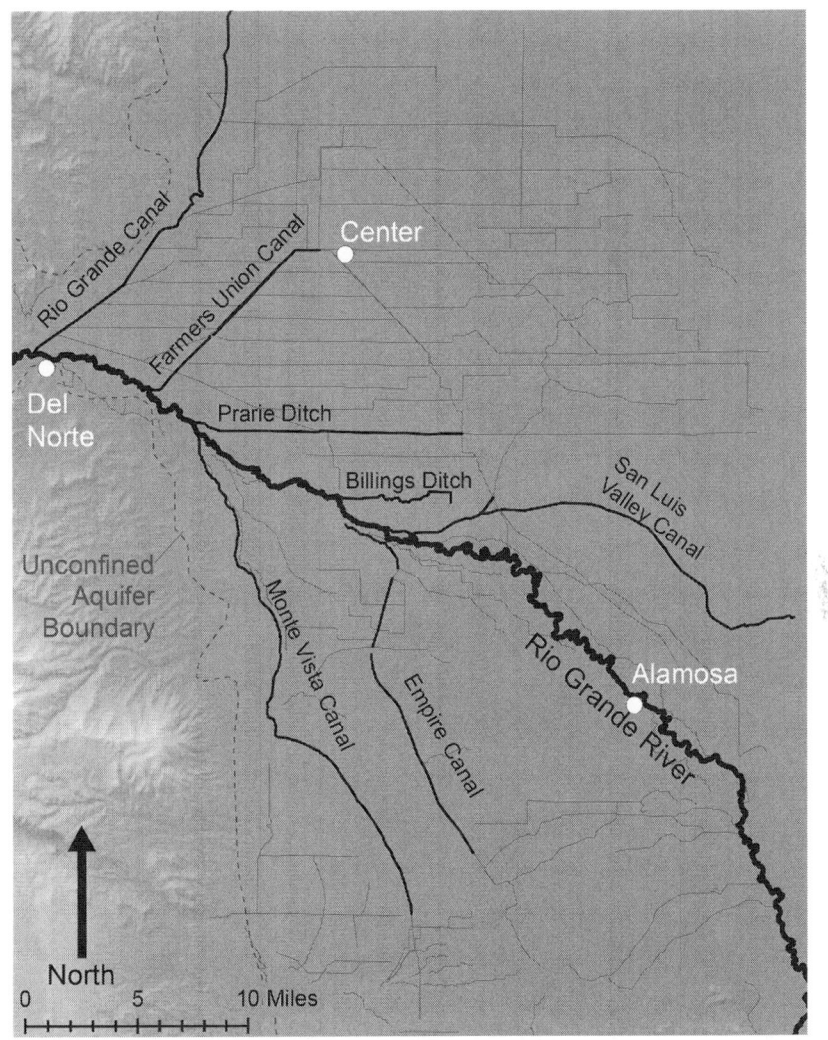

Map of Primary Irrigation Canals, Laterals, and Ditches in San Luis Valley

San Luis Peoples Ditch

Community of San Luis and Culebra Creek Valley

Rio Grande Canal Diversion Dam on the Rio Grande

Rio Grande Canal at Headgate

San Luis Valley Irrigation District Office in Center, Colorado

Rio Grande at Southern End of San Luis Valley

Rio Grande Reservoir

Closed Basin Drain and San Luis Valley

Center Pivot Sprinkler, San Luis Valley

Alfalfa Stacks, San Luis Valley

Water for Sale, San Luis Valley

CRISIS IN THE VALLEY

We've installed all the efficiencies. We use every drop of water
consumptively through these sprinklers. There's a little tiny
bit of return flow, but not much. And we have, in my personal
opinion, exceeded our ability to sustain the system that we
built in good times.

— Steve Vandiver (2013)

On Alamosa's eastern edge, a coffee-brown building with an asymmetrical roof sits on the south side of U.S. Highway 160. Its primary occupant is the Closed Basin Division, San Luis Valley Project, U.S. Bureau of Reclamation, although it also houses the Rio Grande Water Conservation District. Inside, partway down a quiet hallway, a copious, well-lit office provides the professional space of Steve Vandiver, the district's general manager.

Vandiver could pass for a young Wilford Brimley, the well-known character actor. His size suggests that he could make a decent living wrestling bears in a circus, in which case one might be wise to bet on Vandiver. His deep voice rolls out from beneath a white mustache that hides an impish grin seemingly more at home on a man half his size. His desktop is draped with papers and files. A keyboard on his right is backed by two computer screens covered with sticky notes. Where not obscured by bookshelves and stuffed cabinets, the walls are covered with pictures of family and fishing exploits. A Big Gulp is parked by his right elbow and an insulated water bottle sits diagonally across the desk. Personal hydration is presumably important.

Vandiver has been engaged with the Rio Grande and water and water rights in the San Luis Valley for more than forty years. Until 2005, he was the Division 3 Engineer, the chief administrator of water rights in Colorado's Rio Grande watershed. In that role, through droughts and wet spells, for twenty-four years, he consistently failed to make everyone happy, though making everyone happy admittedly was not part of his job description. In the San Luis Valley, there is never enough water to satisfy all parties. More recently, as general manager of the Rio Grande Water Conservation District, a position arguably with greater visibility but less authority, he is still unlikely to please everybody. The district was created in 1967 by the Colorado legislature and the valley's voters to protect, enhance, and develop the valley's water resources. Since then it has morphed into a clearinghouse for most of the valley's water issues that do not fall to the Division 3 Engineer. As clearing-houses go, it is noisy, with many voices.

After collecting an engineering degree from the University of Colorado, Vandiver "escaped," in his words, to the San Luis Valley. He never left. When it comes to water, Vandiver has strong opinions and rarely hesitates to state them. He acknowledges when they are his and not those of his board of directors, although for someone with his longevity and credibility in the valley he may be forgiven if, to many valley residents, the distinction is fuzzy. He admits a preference for the good ol' days, however defined, and makes no apology for believing that, when it comes to water, agriculture is a higher calling than just about everything else, especially urban use. "I don't personally believe that just growing cities bigger is the best use for our water." He pauses, then adds, "That's my personal opinion."[1]

Describing the San Luis Valley's assets and liabilities, Vandiver is both candid and complimentary.

> It's too cold and it's kind of a Spartan place to live. The wind blows, the dirt blows, and the mosquitoes come, and we have two months of nice weather in September and October and then it's cold as hell. We're not a garden spot, so we're not going to attract a lot of people. We grow particular crops that you can't grow as well anywhere else. Alfalfa in the San Luis Valley is ten times better than across the hill [east of the Front Range], and it has lots more protein for feed value. The alfalfa that comes out of here is tremendous. We grow as good potatoes as anywhere else in the world, including Idaho, just because we have cool nights and warm days and lots of sunshine. It never rains here, so water is applied as it's needed, not when Mother Nature decides it.[2]

Vandiver's current grievance is the lack of water in the Rio Grande. "The trouble we're having right now is this sustained drought. In the best of times we're an over-appropriated system. In the best of times we never have an 'open river,' where there's not a call on the river."[3] By *call*, he refers to a fact of life under Colorado water law guided by prior appropriation. A call occurs when river or stream flow is inadequate to meet all decreed water rights, in which case the holder of the senior water right asks the water commissioner, who works for the division engineer, to *curtail*, or restrict, junior water diversions until his or her senior right is satisfied. A nineteenth-century water commissioner, frequently on horseback, probably carried a penciled list of the water rights and decree dates under his jurisdiction. Over time, he in all probability committed the list to memory even if it had several hundred names and dates. Today, if he or she has not committed the priority list to memory, it is accessible on a laptop computer in his or her truck or with a quick call on a cell phone.

If a call is valid, the commissioner closes the junior's diversion headgate, leaving more water in the stream or river for the aggrieved senior right. A call's effect can extend for miles upstream, including tributaries, all the way to the watershed divide. An experienced commissioner may become sufficiently familiar with his or her stream or river, so as to have a well-honed idea of how many junior rights must be curtailed to satisfy a particular senior right, as well as whether junior rights can be partially satisfied and still meet the more senior obligation. If the sum of all decreed water rights exceeds the actual flow of a stream or river, diversions may render the stream or river completely dry, a condition obviously adverse to healthy fisheries or recreational boating. This feature of prior appropriation explains why stretches of many Colorado streams and some larger rivers are dry by midsummer, not because the mountain snowfields have melted or because summer rains have failed to arrive, but because the streams and rivers have been completely diverted for irrigation.

Vandiver adds, "If you look at the amount of water that was appropriated by 1903, that covers the typical flows of the river, so basically, by the turn of that century enough had been appropriated where there was a call all the time. In forty years here, I've seen an *open river* three times."[4] An open river exists when there are no calls in effect, typically during spring floods when the river is flowing full, reservoirs are brimming, and Colorado is in compliance with the Rio Grande Compact. During Vandiver's tenure, just three times in forty years, for about ten days at a time, the Rio Grande had enough water to satisfy all San Luis Valley demands.

Farmers and ranchers follow snowpack data as diligently as they monitor grain and cattle prices and their children's grades. Their groceries and mortgage payments, their livelihoods, depend on the snowpack that has accumulated in the mountains by winter's end and the resulting river and stream flows. When encountering a neighbor at the hardware store or while checking on adjacent irrigation ditches, the conversation is as apt to be about the snowpack as it is the latest Congressional folly. By April or May, they have a good idea—barring natural calamities like summer hailstorms or insect plagues—what kind of year they might have. Yet being aware of how much snow is in the mountains does not put water in ditches and canals. Knowing what is happening does not equate with a capacity to control it. Knowing provides only an estimate of how much or how little water you will have to work with that year. In wet years and dry, western farmers and ranchers stick it out. Snowpack is directly or indirectly the source of most irrigation water in Colorado. If it does not snow this year, the belief is, it surely will next year. The West's farmers and ranchers are among its most stubborn yet optimistic inhabitants.

But it is not all about snowpack. Since the early 1970s, the increased use of center pivot sprinkler irrigation and the resulting improvement in irrigation efficiency have created a double-edged sword. As center pivot irrigation became more widely used, irrigators also drew increasingly upon the valley's prodigious subsurface reservoir, pulling water from the unconfined aquifer like someone drawing from a bank savings account. In this analogy, however, no one knew with any certainty how much money was in the account, while nearly everyone claimed an equal right to it.

Farmers monitored their crops, calculated how much water was needed and when, and then applied it with the flip of an electrical switch to activate their sprinkler system. There was little or no waste; almost no water percolated deeper than the root zone, resulting in very little aquifer recharge during the irrigation season. Total consumptive use by irrigating primarily with groundwater and sprinklers resulted in an irrigation efficiency approaching 75 percent by the beginning of the twenty-first century.[5] Water applied to the crop is consumed by evapotranspiration while return flow is practically nonexistent. Crop yields increased while aquifer recharge decreased.

Nature challenged the valley's cherished status quo beginning late in 2001, when a new drought settled over the region. The American Southwest experienced less snowfall in winter, less rainfall in summer, and lower river flows. Alpine snowfields withered and mountain streams became trickles. At Del Norte, the Rio Grande's total flow for Water Year 2002 was 153,800 AF, a paltry 24 percent of the 1890–2015 average.[6] Even the Dust Bowl years of the 1930s yielded a greater average annual runoff of 593,600 AF. And if the Rio Grande's flow decreased, so did those of the mountain tributaries emerging into the San Luis Valley. Irrigators who relied on surface water were hurt first. Severe water shortages were ruinous, especially for hay meadows at the canyon mouths surrounding the valley. Where streams typically flowed vigorously each spring, their channels became ribbons of rock and sand.

A more ominous threat was the steady decline in groundwater levels in the heavily used unconfined aquifer, which depended on leaky irrigation canals, less efficient irrigation, and voluntary recharge efforts. *Conjunctive use*, the coordinated use of surface streams and hydraulically connected groundwater aquifers, supported the sprawling center pivot irrigation systems that underpinned the valley's alfalfa, barley, and potato crops. As groundwater levels continued to drop the relationships between the drought, the Rio Grande, and the groundwater, so many of those elements relied upon became the central topic of discussion throughout the valley.

Expansion of center pivot irrigation over the past fifty years has resulted in increased consumptive use of water in the valley. Irrigated acreage increased from approximately 400,000 acres in 1950 to about 600,000 acres in 2004, while consumptive use of water increased from about 550,000 AF to 850,000 AF. Much of the increase in consumptive use was provided by groundwater, especially since the early 1990s.[7] Groundwater pumping has more than tripled since 1950, while surface water diversions have varied between several peaks of 1.6 million AF to just 300,000 AF in 2002 in response to climate and river fluctuations. The latter was the lowest on record between 1950 and 2002.[8]

The Rio Grande Water Conservation District has monitored the valley's unconfined aquifer since 1976. During monitoring's first years, the amount of stored groundwater in the aquifer decreased by about 300,000 AF. In subsequent years, stored groundwater experienced both gains and losses as Rio Grande runoff and agricultural withdrawals fluctuated. But between the end of 2001 and late summer 2002—a period which covers the 2002 growing season during severe drought—the aquifer experienced a loss of over 400,000 AF. After 2010, in response to continuing drought and largely unchanged irrigation practices, it declined steadily. By mid-2013 the aquifer had experienced a loss of an additional 500,000 AF, or a total loss of groundwater storage of more than 1.3 million AF since 1976. Since that historic low in 2013, stored groundwater has recovered several hundred thousand AF, but nonetheless declined by almost as much during the extremely dry 2018 growing season.[9] Perhaps the greatest worry was that no one knew just where the bottom of the straw was, how much deeper the valley's unconfined aquifer could be drawn down without hearing a figurative slurping sound. If the historically prodigious snows did not return to the mountains, if the rains did not come, if the drought did not break, if the Rio Grande did not recover, the future looked bleak.

The San Luis Valley's historic wetlands also began to dry out. Wetlands in low areas, such as sloughs adjacent to the river, suffered, while others, where runoff collected from irrigated fields, also dried up. The drought convinced many that the uncertainties of climate change posed equal or greater risks to the valley's agricultural economy than had AWDI's export scheme several decades earlier.

Worse, the drought pitted surface water users against groundwater users. The fact that conjunctive use was even possible convinced surface water users that declining groundwater levels were connected to the water missing from their canals and ditches. Surface water users had been first in the valley, first to irrigate, and generally had senior water rights, many dating to the 1870s, yet they were the first to see their water disappear.

To a point, the complainers were correct. As groundwater levels declined, the rate at which water seeped from the canals and ditches increased to a hydraulic maximum, rendering irrigators reliant on surface water even more anxious. Meanwhile, many groundwater users with junior water rights continued pumping and irrigating. The prior appropriation system was being turned on its head. Neighbors began to criticize neighbors. Surface water users wanted the division engineer to order the offending irrigation wells turned off. The critical question, however, was which wells? Some argued that all wells with junior rights be turned off. But knowledge of the valley's subsurface hydrology was insufficient. The division engineer could not identify with any certainty which of the hundreds of wells impacted which surface water rights. Moreover, the state engineer had never promulgated rules by which offending wells could be ordered to cease pumping or their owners forced to prepare an *augmentation plan*, which is "a court-approved plan whereby a party seeking to divert water out of priority provides replacement water or augmentation water to the affected river in an amount necessary to prevent injury to other water users."[10] An approved plan replaces water in the timing, place, and amount necessary to compensate the injured senior water rights.

Augmentation plans were authorized by Colorado's Water Rights Determination and Administration Act of 1969. Once approved by a water court, an executed plan allows out-of-priority use. Without appropriate rules and augmentation plans in force across the valley floor, the Division 3 Engineer could not act. A stalemate ensued. The same 1969 statute also requires all groundwater wells to be adjudicated in water courts, thereby integrating surface and groundwater rights. It was a new wrinkle in Colorado water law but has been used successfully ever since its legislative approval.

Underpinning all this was the fact that decreed water rights to the Rio Grande total approximately 7,000 to 8,000 CFS, far more than the river flows in all but the most generous and infrequent spring floods. If Colorado had not chosen to be a willing partner to the Rio Grande Compact and pass a portion of the river's flow to New Mexico, there is an excellent likelihood that the San Luis Valley would not allow a drop of the river to leave the state. Various individuals were asked the same question while this book was researched: "If the Rio Grande Compact did not require river water be passed on to downriver states, would San Luis Valley farmers and ranchers still be draining the river as they did in the 1890s?" Those queried included irrigators, environmentalists, water rights experts, federal and state regulatory agents, fishing recreationalists, and retirees with significant regional experience and knowledge. Without exception, they responded "yes." A separate investigation of the river and associated irrigation practices concluded similarly: "Without the need to meet Compact deliveries the reaches of the river below the major diversions could be dewatered for a portion of most years. Compact requirements are the only thing that keeps the river flowing below Alamosa in most years."[11]

◆   ◆   ◆

Within the realm of Colorado water law, the link between agricultural *consumptive use* and *beneficial use* is important yet fuzzy. The former is a measure of how much water a plant uses, or consumes, through evapotranspiration, plus evaporation losses from the irrigated field, while the latter is a less specific though very meaningful legal term, defined in Colorado statutes as "that amount of water that is reasonable and appropriate under reasonably efficient practices to accomplish without waste the purpose for which the appropriation is lawfully made."[12] Where consumptive use is a technical agronomic or hydrological term that expresses a quantity of water, its legal application in Colorado plays an important role when water uses are being transferred. The latter is a legal definition burdened with Colorado's water history. In practice they are related. Consumptive use is how beneficial use is quantified.

Unfortunately, almost no one measures either one. Consumptive use is very difficult to accurately measure in the field and is only calculated when a change-of-use case is presented to a district water court. A "change case," as it is called, occurs when a water right is being transferred between parties; for example, when a farmer sells his water rights to a suburban developer and the beneficial use will be changed from irrigation to domestic or municipal. This is important because a water right is limited to the beneficial use—in practice, consumptive use—not the amount of water that is diverted at a headgate. The balance of the diverted amount of water, typically referred to as transit loss, is not part of a water right, nor is return flow. As succinctly expounded by former Colorado Supreme Court Justice Gregory Hobbs, "Beneficial use is the basis, measure, and limit of a water right."[13]

The legal process involved in transferring an agricultural water right and changing its use requires an historical accounting of what was planted and when and how it was irrigated over a preferred period of twenty-five years.[14] From this information the water court judge, with input from the parties and their experts, determines how much water was consumptively used during the period of use and thus how much water is available for a different use.

Agricultural consumptive use estimates clearly depend on how water is applied and what is grown. Sprinklers are more efficient and can do more with less water than can traditional furrow or flood irrigation. Twenty-first-century crop yields are greater than those possible in the late nineteenth and early twentieth centuries when many water rights were originally established. And increased crop yields require more water. More water is needed to grow three tons of alfalfa than two tons. If an irrigator has over time inadvertently or intentionally increased his or her consumptive use, that increase is wrapped into the twenty-five-year average for any prospective change case before a water court. If there is no change case being heard and consumptive use has been increased without challenge by a division or state engineer, or water court—which almost never happens—the irrigator benefits. The reverse may also be true. If an irrigator changes his or her crop mix to favor grains while decreasing water-loving crops like alfalfa, his or her consumptive use will normally be less. Grains require less water to grow than alfalfa. Likewise, if an irrigator abandons his efficient center-pivot sprinklers and returns to furrow irrigation—a possible yet unlikely occurrence—there is a marked probability that total consumptive use would decline, and return flow to the receiving stream or aquifer would also increase.

The reality is even murkier. The consequences of one or two farmers exploiting a less-than-perfect system might not show up in a valley aquifer's annual water budget or among a string of streamside irrigated fields. It could be calculated, but not with great accuracy. Groundwater flow and aquifer storage science accept imprecision and limiting assumptions. Water engineers and farmers do the best they can within practical limits because the measurement of the various factors is imprecise. But if hundreds of farmers operating on tens of thousands of acres exploit the same imperfect system over decades, it is difficult to avoid the conclusion that they will increase their consumptive use. Is this an inappropriate enhancement of a water right? Past and present professionals of the State Engineer's Office have a term for it: "legal expansion."[15]

Few would argue that efficient irrigation of valuable food or forage crops is not a beneficial use under Colorado law. But which measure of beneficial use applies? The amount of water consumptively used by the San Luis Valley's many irrigators has certainly increased over time just by the nature of changes in the crops under irrigation and how they are watered. This was especially true following the transition to center pivot sprinklers over the past half-century, and also as alfalfa, a water consumer of the first order, became profitable to grow.

Improvements in crop genetics over the past century have likewise played a role in consumptive use. And climate change has contributed, as longer growing seasons result in increased crop water consumption. Increases in consumptive use almost certainly occurred with the historical transition from flood irrigation to subirrigation, and then from subirrigation to overhead sprinklers. With each improvement in application efficiency came an increase in consumptive use and a decrease in return flow, both being contemporaneous with increased withdrawals from the same immense San Luis Valley aquifer used in common. Everyone is in it together.

Changes in consumptive use may have gradually and incrementally enhanced water rights in a manner never considered or addressed by Colorado's legislature, water courts, a state engineer, or even the state's widespread and influential agricultural community, and one might wonder why state or division engineers, or even the Colorado legislature, have failed to address this issue. It may be that the engineers and legislators want nothing to do with forcing the application of theoretical limits to consumptive use any more than they want to tell farmers what crops to grow. As long as diverted quantities remain unchanged, little can be done about it. The long-term losers are junior water rights and surface and groundwater bodies subject to decreased return flows.

Meanwhile, in the San Luis Valley, an increased understanding of the valley's hydrology, in combination with more efficient irrigation, has led to the fine-tuning of conjunctive use. The valley's aquifer provides needed storage, enabling irrigators to squeeze ever more productivity from the Rio Grande and valley floor. When drought appeared in 2001 and stayed through 2002 and beyond, however, the combination of reduced river and stream flows, a declining aquifer, increased consumptive use, and irrigation pumps that could not be turned off resulted in the entire system in slow-motion freefall.

CHAPTER TWELVE

COLLABORATION AS AN ECONOMIC AND CULTURAL TOOL

The best thing going in the valley is the sense of community. What affects one of us affects all of us. Our geography, topography makes a sense of community, because you can see from east to west and you can see from north to south.

— Ralph Curtis (2013)

"We're still 65 percent of normal snowpack and runoff, and we're struggling aquifer-wise. We reduced pumping by about 30 percent over the last two years and we're still dropping the aquifer."[1] Spring runoff for 2014 had begun, and Steve Vandiver was lamenting the ongoing drought and how it was affecting the valley's primary aquifer. Without a healthy spring runoff from the Rio Grande, it made little difference whether irrigators diverted water from their canals and ditches or if they pumped groundwater from the aquifer. The river feeds the canals that recharge the aquifer. It is the same water. The significant difference is in the timing: surface runoff without a reservoir must be diverted as it flows past the headgate. Groundwater beneath your feet can be used when needed.

Vandiver's desk remained a study in deliberate chaos. The Big Gulp cup was in the same place. Notes were still stuck to his computer monitor. He spends most of his working hours at his desk. Someone with his knowledge, experience, and responsibility expects to communicate regularly with regulators, agency heads, irrigators, journalists, and environmentalists, answering their questions and explaining his agency's actions, as well as leading the planning and implementation of the valley's aquifer recovery efforts. For questions about what is wrong or right with water use in the San Luis Valley, Vandiver was still the go-to guy.

Nineteenth- and early twentieth-century flood irrigation created the aquifer, and now the valley's agricultural economy depends upon its ability to sustainably receive and yield water. But its sustainability is in question, not just because of the drought, but also because of the decreased return flows due to sprinkler irrigation efficiency. Nearly one thousand square miles of the valley were devoted to irrigation in 1998, two-thirds of that to alfalfa and grass hay.[2] Alfalfa is the valley's principal water consumer; the valley's irrigated grass hay meadows consume nearly as much. While this may seem a shortsighted allegiance to a single water-consuming crop, the valley's agricultural economy annually accounts for $285 million in crops and $40 million in livestock.[3] More important to many valley residents is that agriculture is a way of life. Cognizant of all these facts, Vandiver makes no apology for doing what he can, ethically and legally, to support agriculture in the San Luis Valley.

Between 2002 and 2014, of all the major watersheds in Colorado, the Rio Grande was alone in experiencing almost continuous drought. As the Rio Grande goes, so goes the aquifer—and the valley's economy, which according to Vandiver teeters on disaster. "We're not getting the inflows, we're not getting the diversions to recharge the aquifer, and so it [groundwater levels in the aquifer] continue to go down. We developed this water situation in the '70s, '80s and '90s, and then [the climate] changed. Now what do you do? How do you unscramble the egg?"[4]

The valley's love affair with groundwater began after valley farmers spent decades practicing and perfecting flood irrigation and subirrigation. Prior to construction of the valley's major irrigation canals in the 1880s, depth to groundwater could have been measured in tens of feet. Flood irrigation and large-scale subirrigation before 1900 raised groundwater levels in some places to within several feet of the land surface. Eventually, the new aquifer made itself obvious and thousands of groundwater wells were completed, most less than several hundred feet deep.[5]

Before the latest drought commenced in late 2001 and 2002, however, the beginning of what would become a different festering conflict began in the early 1950s, in the north end of the valley along lower Saguache Creek, where irrigators who had been relying solely on surface irrigation for decades objected to recent arrivals completing wells and using groundwater. Surface flows began to decline. The pattern repeated itself along upper San Luis Creek in the early to mid-1970s after center pivot sprinklers were introduced. The importance of ground-water storage became even more pronounced when farmers realized that overhead sprinklers allowed them to grow four tons of alfalfa per acre each year, instead of two, and that they made more money with alfalfa than with a single crop of barley. But since at least 1980, as the number of center pivots increased across the valley, more and more irrigators realized that increased reliance on groundwater resulted in less water in canals and ditches. To them, the close relationship between surface flows and the underlying aquifer was obvious. Why, they subsequently asked, did the Division 3 Engineer not order offending well pumps turned off? After all, most wells were junior to most surface water rights. According to Vandiver, the answer was obvious. "Back when they first started drilling wells in this valley, there was no recognition that there was any impact on anything. The science hadn't caught up with what the impact to streams was going to be."[6]

The science of groundwater hydraulics was in its infancy a century ago, especially in rural Colorado. No more aware were the legal profession and Colorado's water courts, which had little grasp of how to deal with groundwater and its mysterious and invisible relationship with surface water. This lack of knowledge was more widespread than just in Colorado. In 1861, the Ohio Supreme Court had thrown up its hands over groundwater and how it might be administered: "the existence, origin, movement and course of such [ground] waters, and the causes which govern and direct their movements, are so secret, occult and concealed, that an attempt to administer any set of legal rules in respect to them would be involved in hopeless uncertainty and would be therefore, practically impossible."[7]

Groundwater engineers and geologists chuckle at this statement today. Even a century after the Ohio Supreme Court's ruling, the situation was not much clearer in the San Luis Valley. But given Colorado's long experience with prior appropriation, how did it get so out of whack? One of the chief advantages of the prior appropriation system is that it provides predictability about the availability of water when supplies are scant. Users know their place in line; they should be able to adapt. How did junior groundwater appropriators effectively leap ahead of senior surface water rights?

Prior to 1960, there was no long-term monitoring of groundwater levels or artesian pressures in San Luis Valley aquifers. Colorado did not regulate groundwater wells until 1965, and not until four years later did the legislature mandate the integration of surface and groundwater uses into what is known as *conjunctive use* and require this new policy to be incorporated into Colorado's prior-appropriation law. The state engineer did not, and could not, begin to integrate groundwater rights with decreed surface water rights until after 1969. Given the markedly different eras when Colorado began to address surface and groundwater rights, a degree of flexibility would seem to have been in order, and today the germane questions are, first, whether the two sets of rights have since been integrated, and second, is prior appropriation capable of working fairly when the two sets of rights are merged?

Vandiver agreed that integrating the two systems was difficult and conceded that failure to do so "let 1970 wells run when you've got an 1870 surface right that's being curtailed."[8] The ideal response may have been for the Division 3 Engineer to order out-of-priority wells shut down unless and until the owner could provide augmentation water to the affected party, but no one in the valley possessed any certitude at the time about which well or wells were affecting which surface water rights, a problem that would not begin to be addressed for almost another thirty years. The state's adoption of conjunctive use in 1969, while necessary and important, created statewide confusion. With periodic drought playing an intermittent and complicating role, the state engineer and his agent in the San Luis Valley, the Division 3 Engineer, were stumped.[9]

The situation lurched forward. By the mid-1980s, surface water rights were being *curtailed*—forced to reduce otherwise legal diversions—to help Colorado meet the Rio Grande flow obligations to New Mexico. The U.S. Bureau of Reclamation attempted to help satisfy the state's compact obligation in 1985 by constructing the Closed Basin Drain and channeling pumped groundwater to the Rio Grande and on to New Mexico, but the project was unable to deliver more than 43,000 AF in a year, when it was originally designed to annually produce more than twice that amount.[10]

The problem festered, subsided, and rose again, usually in response to variable annual runoff. Some senior surface rights owners wanted *all* groundwater pumps turned off until every senior surface water right was satisfied. In a perfect world, with a fully integrated prior-appropriation system in place, this may have been enforceable. However, the state engineer had failed to promulgate rules by which he could identify offending wells and order them turned off. A 1975 effort to write these rules was quashed by the courts when the rules were deemed too complex to be fairly administered without threatening water rights of either type.

By the late 1990s, parties to the continuing kerfuffle finally agreed on one thing: they needed more information. The valley's web of leaking canals and ditches overlying a layered, two-aquifer groundwater system was a myriad of partially understood generalities. No one at the time adequately understood the valley's hydrology, and until that shortcoming was corrected no one could be certain which well or wells, if any, affected any specific ditch or canal. And without legally enforceable rules, the division engineer lacked authority to do much about it.

With broad support from the valley's water community and some serious lobbying in Denver, the 1998 state legislature passed House Bill 98–1011, which directed the state engineer and the Colorado Water Conservation Board (CWCB) to study the relationship among the valley's aquifers, surface streams, irrigation canals, and irrigation methods. Carried out between 1998 and 2004, the study established the Rio Grande Decision Support System (RGDSS) to provide the technical underpinning for all ensuing groundwater development and water rights administration in the San Luis Valley. Given that the valley's annual sum of surface water and groundwater inflows and outflows approximated *1 million* AF, the importance of the new support system to agricultural water users could hardly be overestimated.[11]

To achieve this needed understanding, the state engineer and CWCB agreed that a computer model of the valley's hydrologic system had to be created and calibrated, and the relationship between the surface and groundwater systems had to be quantified. The model also had to be able to account for irrigation return flows to the Rio Grande, because return flows, especially during winter months, provide 20 to 25 percent of the state's annual compact obligation to downriver states and Mexico.[12] Development of the computer model commenced, but the complexity and difficulty of the task made its completion a lengthy process. Not until 2014 could the model predict with reasonable accuracy the movement of surface and groundwater in a manner useful to the valley's water managers.

While all this was ongoing, San Luis Valley irrigators did not cease farming. In 2002, when drought refocused everyone's attention, the lack of surface runoff led well owners to increase their groundwater use to record levels. *Net groundwater consumptive uses* in the valley that year were more than twice the amount that the Rio Grande discharged to the valley,[13] a quantity that could only result from drawing upon water stored in the shallow aquifer. Meanwhile, the valley's irrigators continued arguing over whose water rights should be curtailed.

Around the same time, and before the computer model was functional and the state engineer and CWCB could provide relief to the valley's water combatants, Ray Wright, a valley farmer and president of the board of directors of the Rio Grande Water Conservation District, proposed a new approach to integrate the valley's surface water and groundwater rights: creation of *groundwater subdistricts*. He had observed ongoing legal battles and societal friction along the South Platte River in northeastern Colorado and wanted to forestall similar tension in the San Luis Valley. Wright argued that groundwater subdistricts had the potential to solve many long-festering problems. In particular, they would allow the equitable and collective administration of groundwater rights and uses while similarly protecting surface water rights. Subdistricts might also help integrate surface and groundwater claims. Wright's idea was not welcomed by all parties, because it had not been successfully tried elsewhere in Colorado or blessed by the state engineer. And to implement the subdistrict program, the state legislature needed to enact changes to Colorado water law.

Wright's groundwater subdistrict idea was wrapped into Senate Bill 04–222 and presented to the legislature in the spring of 2004. Testifying before the legislative body, Wright emphasized that the plan had been developed collaboratively by the Rio Grande Water Conservation District and Rio Grande water users. In addition to admitting that the valley's water users had brought the problems upon themselves, Wright acknowledged that the concept was not a guaranteed fix. Nonetheless, after observing conflicts along the South Platte, Wright and his cohorts feared that the Rio Grande would be next on the state engineer's fix-it list if something innovative was not attempted. And quickly. One thing the valley's water users agreed upon was that solutions applied from outside the valley might prove less desirable than something generated within.[14] Groundwater subdistricts could serve as an alternative to the state engineer regulating individual wells, something many feared would be harmful to one or more well owners, especially in light of the absence of information on how the valley's surface and groundwater resources interacted—in other words, before the results of the RGDSS computer model were available.

No solution to the problem would be cheap or easy, but the new law would have numerous benefits for valley water users.[15] Under Senate

Bill 04–222, groundwater subdistricts could continue to utilize the valley's unconfined aquifer as a reservoir.[16] They would be special taxing districts and allow local interests to manage aquifers in a flexible manner while also meeting the legal requirements of individual augmentation plans.[17] They would be, in effect, geographic communities of common interests established to finance the purchase of augmentation water. Well owners within a subdistrict would pay annual fees based on the amount of groundwater they pumped and the number of acres they irrigated. If they had surface water rights that were used to recharge the aquifer, the amount of water represented by those rights would be credited against the amount of groundwater they pumped. Conceivably, some well owners would pay little or nothing beyond an annual per-irrigated acre fee of seventy-five dollars. The fundamental premise was that, by charging themselves to use groundwater, the number of acres under irrigation would decrease and the unconfined aquifer would stabilize and begin to recover. In exchange for collectively creating a subdistrict water purchasing authority that could buy augmentation water, subdistrict members would be allowed to operate their wells "out of priority." Collected fees would also be used to pay farmers and ranchers to fallow farm acreage—take land out of production.[18] Removing land from production would be another way to reduce total pumping from the aquifer. Likewise, subdistrict user fees could be used to purchase water to help fulfill the state's Rio Grande Compact delivery obligations. And because some potentially affected well owners were suspicious of granting too much authority to a new agency, groundwater subdistricts would have no regulatory authority. They would be purely a financing mechanism. Well owners within a subdistrict boundary would not be forced to join the subdistrict. They would have a choice. They could develop and submit for regulatory and court approval their own augmentation plan, which would require finding a permanent source of water to compensate injured surface water rights. Or they could join the subdistrict. Or they could shut down their well. It was their decision. Well owners would eventually choose overwhelmingly to join subdistricts.

The legislature approved Senate Bill 04–222, and the governor signed it into law. By that fall the Rio Grande Water Conservation District was preparing to create the valley's first groundwater subdistrict, known formally as Special Improvement District No. 1 of the Rio Grande Water Conservation District, or informally as Subdistrict No. 1. Situated in the San Luis Valley's irrigated center and surrounding the town of Center, it was officially formed in July 2006. Opponents of the concept immediately challenged the legality of Subdistrict No. 1 in court, arguing that *all* wells should be turned off until *all* surface rights were satisfied. Court fights continued for five years, until the Colorado Supreme Court finally confirmed the subdistrict's legal standing and its proposed management plan.

Before it could get fully underway, however, still another approval was required. The subdistrict also needed approval and funding from the U.S. Department of Agriculture and its Conservation Reserve Enhancement Program (CREP). Authorized and funded by Congress, CREP money was another means to compensate irrigators for fallowing their fields and retiring groundwater rights.[19] The concept behind retiring groundwater rights was that getting the aquifer and its agricultural withdrawals back into balance had to be permanent, or else there would be little to keep farmers from overdrawing the aquifer in the future if and when the drought ended and the Rio Grande supplied more water for aquifer recharge and irrigation.

Because of the lengthy court battle over its legality, Subdistrict No. 1 did not begin functioning until 2012. In its initial layout, it covered nearly 174,000 acres, or 272 square miles. More than half the subdistrict's irrigation water was pumped from the unconfined aquifer, so groundwater use was the focus of practically everyone's attention. A stated objective of the subdistrict's formal plan was to reduce irrigated acreage by nearly one-fourth, or 40,000 acres. Another objective was to reestablish the unconfined aquifer's groundwater storage to within 200,000–400,000 AF of 1976 levels within twenty years.[20] Given where the total amount of groundwater storage was in 2012, this would necessitate recovery of an additional 500,000 AF by 2032, while simultaneously managing annual recharge and withdrawals by the irrigation community. Everyone knew it was a tall order even if they had twenty years to achieve it, yet everyone also wanted to avoid abrupt changes that might ruin individual farmers or shock the local economy. The recovery process was to be slow, methodical, and successful.

Improvements showed almost immediately. By the following fall of 2012, groundwater pumping in Subdistrict No. 1 decreased by 27 percent. However, the gain could not be sustained. Over the first two growing seasons (2012 and 2013), water levels in the unconfined aquifer as a whole (of which Subdistrict No. 1 is part) declined to their lowest recorded levels, down by 1.35 million AF since 1976 and by 1.05 million AF since the drought began in 2002.[21] To be truly effective, it was obvious that the groundwater subdistrict approach would have to be extended over broader reaches in the valley. More subdistricts would be required.

Although 2002–2003 were desperately dry years for the Rio Grande, conditions fluctuated over the next fifteen years. The watershed recovered somewhat during 2004–2008, only to experience below-average flows from 2009 through 2013. River yields picked up again over the next four years and remained near average through 2017. Fortunately, as the groundwater subdistrict concept gained momentum, the valley's unconfined aquifer responded favorably. The groundwater management plan and subdistrict concept appeared to be working. Groundwater withdrawals decreased as farmers responded to the new fees that they imposed on themselves. Stated differently, the groundwater subdistrict process forced irrigators to pay closer to the true economic value of water. In accord with traditional economic theory, they became more efficient.

One recurring argument raised by continuing opponents of the groundwater subdistrict regime was that it unfairly increased costs to those farmers using groundwater. But these new fees were really compensating the subdistrict for the purchase of augmentation water, without which individual groundwater users would have been required to provide an augmentation plan or give up using their wells. Nearly everyone within the geographic limits of a subdistrict eventually saw the economy and wisdom of signing up. Some acceptance might have been grudging, but it occurred.

When it came to consistently decreasing the number of irrigated acres within the subdistrict, irrigators struggled to achieve the 40,000-acre reduction target. In 2013, farmers fallowed about 9,000 acres, either through negotiated agreements with the subdistrict or by enrolling in one of two federal CREP programs.[22] (One CREP program paid farmers $175 per acre to permanently retire farm acreage. Another program allowed farmers to retire acreage for 15 years, paying the same amount per acre.[23]) Such a fee per acre is paltry compared to what a farmer might earn per acre in an average year, even after accounting for seed, fertilizer, fuel, electricity, and labor costs. Following the subdistrict's formation in 2012, both CREP programs initially removed only 6,000 to 7,000 acres from production.[24] Over its first four years, the subdistrict achieved less than 30 percent of its acreage reduction goal. Some farmers decided to fallow marginal land without outside compensation, deciding that avoiding the costs of production would save them more than they would gain by signing up for either CREP program, while providing them the flexibility to put land back into production when commodity prices rise or the skies open and Rio Grande flows return to historical norms.

Other difficulties also arose. While the subdistrict's groundwater withdrawals were slowly decreasing, the broader San Luis Valley was experiencing what David Robbins, a Denver-based attorney representing the Rio Grande Water Conservation District, later described as a "perfect storm."[25] Modern farmers are more than workers on the land. To be successful, to remain in business, they must be shrewd businesspeople. Rising farm commodity prices occasionally present business opportunities, and astute farmers typically respond by planting different crops to take advantage of favorable market prices. Even as Subdistrict No. 1 was being created, opportunity arose in the alfalfa market. Grains and especially potatoes had been the valley's staple crops for years, but the development of large commercial dairy operations in nearby New Mexico and Texas increased the regional demand, and thus prices, for quality alfalfa. San Luis Valley farmers responded by planting less grain and potatoes and more alfalfa.[26]

Alfalfa, like grass hay, is a forage crop. Compared to grass hay, however, alfalfa has more protein, a desirable quality to dairy cattle herd owners. Research agronomists have scientifically fine-tuned alfalfa over the years to enhance its palatability to livestock and increase its nutritional value, resulting in its preference over grass hay by most farmers. In spite of its altitude and short growing season, the San Luis Valley allows farmers to grow at least two and sometimes four cuttings in a season. With a reliable nearby market willing to pay premium prices for quality alfalfa, farmers responded logically.

The flip side of the farmers' decision to increase alfalfa production led inescapably to increased water consumption. Fourteen to twenty inches of water is generally adequate to grow a crop of grain or potatoes in the San Luis Valley, while alfalfa can require thirty to thirty-six inches or more per growing season. Subdistrict No. 1 demonstrated its capability to reduce groundwater pumping, but the high prices being paid for alfalfa inharmoniously discouraged other well owners in the valley from reducing their groundwater consumption.[27] Farmers thus had a perverse financial incentive to *increase* their use of groundwater to grow alfalfa even while they were planning more groundwater subdistricts in order to reduce total groundwater pumping. Between 2000 and 2016, approximately 23,000 acres were voluntarily removed from production against the initial target of 40,000 acres.[28] By 2019, however, the district backslid to having removed only 10,000 acres.[29] An obvious inference to be taken from this story and these numbers is that Subdistrict No. 1 groundwater users varied their annual uses, in response to both commodity markets and available water and their collective commitment to aquifer recovery.

Beginning in 2008, the state engineer required all well owners to install meters and annually report how much water they pumped from each well. The submitted information allowed the State Engineer's Office to develop historical records, calibrate and fine-tune the valley's computer model, and compare the quantity of groundwater actually pumped against the associated water right. There were initial concerns among some well owners that the state engineer might use this new information to limit annual pumping to something like a five-year average. This eventuality has not come to pass, and there is a question whether the state engineer even has such authority.[30]

Another important factor that has been recognized is the Office of the State Engineer's mandate that the aquifers in the San Luis Valley be managed "sustainably." The State Engineer, however, was merely citing Senate Bill 04–222: "Use of the confined and unconfined aquifers shall be regulated so as to maintain a sustainable water supply in each aquifer system, with due regard for the daily, seasonal, and long-term demand for underground water."[31] The mandate to manage groundwater resources sustainably was more than a subjective regulatory edict—it was state law. Fortunately, San Luis Valley irrigators generally agree with this requirement. To deny it would provide an opportunity for future irrigators to simply drain the aquifer in pursuit of maximum production and, presumably, profit, then walk away from the valley. Sustainability appears to be a serious effort by the State of Colorado to conserve the valley's aquifers in perpetuity.

Altogether, these factors continue to make for an increasingly complex farming environment. Water managers across the valley foresee a strange future and can only shake their heads in wonderment and push on. Yet despite these variables, the San Luis Valley's unconfined aquifer situation has improved since implementation of groundwater subdistricts. Compared to the greatest groundwater withdrawals that occurred in 2013, by early 2018, metered withdrawals had decreased by about one-third.[32] Obviously, though, meeting the aquifer's recovery goal will depend upon future flows of the Rio Grande as much as the agricultural community's willingness to self-limit its dependence on groundwater.

• • •

As drought, irrigation technology, litigation, and regulatory moves and countermoves affected the San Luis Valley over the past decades, the acequia culture of the Spanish-American farmers of Costilla and Conejos Counties in the valley's south end continue largely as they have since 1852. Their ancestors were the first to settle and irrigate lands along Culebra Creek, and their descendants continue to live on and work the land in the twenty-first century. Their community-based cultural precept of sharing water was largely unaffected by early Colorado water law after the Colorado Territorial Legislature authorized their distinctive differences prior to statehood. Following statehood and adoption of prior appropriation as the guiding principle of water use, the Spanish-American farmers in the valley's south end participated in the adjudication system and obtained what today are among the most senior water rights in the state.

In spite of acequia culture's existence in the San Luis Valley well before Colorado statehood, it was not until state legislative passage of the 2009 Acequia Recognition Law (amended in 2013) that Colorado allowed acequias to legally continue their traditional roles in controlling and governing community access to water. Acequia associations, long recognized by their communities, are now recognized by the state in Costilla and Conejos and two additional southern Colorado counties if they follow specified organizational and management procedures. Acequias thus possess legal sanction for many of the practices they have long implemented, including many forms of water sharing and procedural requirements limiting the ability of members to transfer water rights out of the acequia.[33]

During his years as the Division 3 Engineer, Steve Vandiver was called upon to incorporate acequia culture into modern Colorado water law administration, and today acequias are governed the same as other ditches and canals. To Vandiver, contemporary acequias are "a distinction without a difference from the other ditches... There's no sharing of the shortages amongst the ditches on a particular stream. It's [the acequias are] administered in priority. [But] what they do internally, we don't get in any ditch company's business, whoever they are."[34] Acequia diversions from a stream or river are controlled by a state water commissioner, just as they are elsewhere in Colorado. The headgate is the point of diversion regulated by law. Along Culebra Creek, the water commissioner's responsibilities also end at the headgate, where he or she shows up to adjust headgate flow.[35]

On the farm side of the headgate, however, historic acequia culture prevails. Each acequia, or ditch, has a legal decree, but the owners of the acequia agree among themselves how and when to divide the water. The divisions are administered by a hired mayordomo. Today's owners prefer the term *herederos* to describe themselves, meaning heirs, to the historic idiom *parciantes*. Many herederos connect their ownership to the valley's first settlers of the 1850s. Their children may leave the valley, perhaps to attend college or acquire technical education, although many eventually return. Older members still speak Spanish among themselves, leaving English to the young. The acequia's annual spring cleaning still occurs, and each owner is expected to participate or hire someone to assist in his or her place. Even their crop and irrigation preferences are traditional; native or traditional garden plant varieties are commonly grown. Modern sprinkler irrigation is used in the valley's south end, although many herederos still employ flood or furrow methods. Many of the valley's Spanish-American farmers have modified their irrigation culture relatively little since their ancestors arrived, especially when compared to irrigation practices elsewhere in the valley. Nevertheless, their practices constitute a minor use of water in the valley. There is no reason to suspect they will change anytime soon.

CHAPTER THIRTEEN

ELSEWHERE IN COLORADO AND THE WEST

*Virtually all western rivers of any significant size have been
altered to enable human control and use of their water.
The result is a complex network of diversion dams, storage
dams, water-delivery canals, ditches, laterals, siphons, and
drainage systems.*

— Lawrence J. MacDonnell, *From Reclamation to Sustainability* (1999)

Colorado's agricultural land is distributed unevenly throughout the state. On the plains east of the Rocky Mountains and in most river valleys on the West Slope, wherever there is amenable ground *and* adequate water, farms and crop cultivation occur. Meanwhile, farming in the mountains is difficult to impossible; soils are thin, slopes too steep, and the climate too cold. Not all of Colorado's agricultural land is irrigated. In fact, most is not. About 2.7 million acres out of more than 11 million cropped acres are irrigated.[1] Seventy to 75 percent of Colorado's farmed land relies solely on direct precipitation.

Colorado's most productive and profitable irrigated farmland is in the San Luis Valley and the major river valleys east of the Rockies. Colorado water law presumably is applied uniformly across the state, but local histories and environments differ. So do the results. The San Luis Valley is not Colorado's primary agricultural area. It ranks second, behind the South Platte River Valley, which, in addition to the Arkansas River Valley below Pueblo, is moderately to heavily irrigated. An overview of these two rivers and their agricultural histories is warranted to assess the uniformity of irrigation practices and water rights administration across the state.

◆ ◆ ◆

South Platte River

Like all major rivers originating in Colorado, the South Platte's head-waters are in the mountains. The river heads in South Park, a high valley in the Rockies southwest of Denver. Headwaters streams coalesce and descend through a steep-walled Platte Canyon. From there, the river flows northward through Denver toward Greeley, the region's agricultural center, then bends eastward to Nebraska where it joins the North Platte to form the mainstem of the Platte River.

Prior to the nineteenth century, the South Platte churned across Colorado's eastern plains each spring in a muddy torrent. Few trees crowded its banks; the few copses grew mainly on islands. Above its confluence with the North Platte, it was a quarter-mile wide and easily forded by coach, horse, or even on foot, lending credence to the river's frontier description: "A mile wide and an inch deep." However, the modern South Platte is a fraction of its former self in the spring, and afterward is frequently just a sluggish current between banks covered with cottonwoods, boxelder, and thickets of brush, its size and habit due almost entirely to irrigation withdrawals.

The South Platte played a major role in Colorado's gold strike. The first immigrants lived in clusters of cabins at the confluence of Cherry Creek and the river before the clusters fused into what became Denver. The young city used the river for domestic and municipal water and to irrigate small gardens and vegetable plots,[2] but almost immediately, "Agricultural irrigation systems began to transform the land around the fledgling city from dry prairie into productive farmland, providing fresh food for residents of the Front Range."[3] Irrigation was an arduous undertaking for the valley's Euro-American farmers; they were forced to adapt just as their brethren did in the San Luis Valley. Denver newspaper editorials from the time predicted that agriculture would ultimately prove more profitable than mining, even if irrigation were required.[4]

Irrigated agriculture also took root along the Cache la Poudre River, a major tributary to the South Platte that heads in the mountains northwest of Denver and joins the South Platte near Greeley. Euro-American settlers began diverting water for irrigation on the Cache la Poudre as early as the 1860s, and by 1870 were forming cooperatives to develop canals and ditches along the Cache la Poudre, the South Platte, and their tributaries. "The Cache la Poudre Valley in 1882 was declared to be 'one vast network of irrigating canals.'"[5]

Irrigated agriculture became the most important economic activity along the South Platte, the Cache la Poudre, and their primary tributaries well before the twentieth century. Between 1860 and the late 1880s, more than forty major canals were constructed and acquired water rights between Denver and the Nebraska state line. Five canals diverted South Platte water to off-channel storage,[6] and numerous reservoirs were constructed by the 1880s. By 1911, the Cache la Poudre Valley boasted fifteen reservoirs with a combined capacity in excess of 69,000 AF, a volume that then doubled over the next fifteen years.[7] Today the South Platte Valley is the state's most densely irrigated agricultural area, accounting for more than 830,000 irrigated acres, nearly a quarter of Colorado's total. The state's second most heavily irrigated area, the San Luis Valley, has 622,000 irrigated acres, 18 percent of the state total.[8]

The combination of the native South Platte runoff and twenty-seven transbasin diversions considerably increase the total amount of water flowing in the valley. The South Platte's native annual runoff averages about 1.5 million AF,[9] but an additional 580,000 AF/YR is transferred to the South Platte watershed by transbasin diversions from west of the Continental Divide. Private companies or state or federal agencies initially underwrote or completed these diversions to serve varied interests, including irrigation, municipalities, and rural water companies. More recently, and importantly, much transbasin water has been redirected from agricultural to municipal uses.

Similar to how Texas, New Mexico, and the Republic of Mexico complained about Colorado's consumption of the Rio Grande in the 1890s, Nebraska protested Colorado's use of the South Platte in 1916. Ultimately reconciled by compact in 1925, the South Platte controversy was complicated by technical arguments over the importance of return flow. Colorado maintained that return flow created a South Platte that was "a stream of constant flow, to and across the interstate line, although its waters have been repeatedly diverted, used, rediverted and reused for irrigation of 1,500,000 acres." The 1,500,000 irrigated acres may have been an exaggeration, but a 1926 report by the Colorado State Agricultural College (eventually renamed Colorado State University, located in Fort Collins) maintained that after Colorado diverted considerable South Platte water, the river's annual flow delivered to Nebraska averaged about 300,000 AF.[10] This was an apparent early twentieth-century attempt to emphasize the magnitude of return flow in sustaining the South Platte's contributions to downriver farmers and Nebraska.

Below its confluence with the Cache la Poudre, the South Platte's present annual flow averages more than 800,000 AF. By the Nebraska state line, irrigation withdrawals have reduced this amount by half.[11] At least six points along the South Platte between Commerce City (near Denver) and the Nebraska state line "dry-up" during most irrigation seasons, typically just below major canal diversions.[12] Return flows typically create a "gaining" stream along much of the river, so the river channel does not remain dry for long.

The South Platte's irrigation regime differs significantly and importantly from that of the Rio Grande in the San Luis Valley. The Rio Grande's flow is largely diverted to the San Luis Valley floor, where return flow recharges the valley's huge unconfined aquifer. Some return flow directly adds to the river, and a portion (e.g., discharge from the Closed Basin Drain) finds its way back to the Rio Grande and continues downriver to other appropriators and to satisfy Colorado's commitment to the Rio Grande Compact. By contrast, numerous canals divert water from the South Platte and its tributaries, but there is no underlying aquifer of comparable size to receive and store return flow for subsequent use by the same farmers, certainly not to a comparable degree as the Rio Grande in the San Luis Valley. Instead, return flow continues down the South Platte Valley either as surface flow to be intercepted by drains, canals, tributaries, and the river, or as groundwater transmitted in the South Platte's much smaller alluvial aquifer. For this reason, South Platte water is diverted, applied, returned, and reused several times before reaching Nebraska, although the quantity decreases down-valley as consumptive use occurs. The distinction between the flow paths and fates of return flow between these two river systems is important. San Luis Valley farmers relying upon groundwater draw from the same aquifer and rely on the same undifferentiated return flow as their neighbors, whereas many South Platte farmers rely on irrigation return flow from up-valley.

One of the most important tweaks to Colorado's priority system came about with passage of the 1969 Water Rights Determination and Administration Act, which permits out-of-priority diversions if a court-approved augmentation plan replaces the lost water. The San Luis Valley's groundwater subdistricts fulfill this function. Within the past few decades, a crisis along the South Platte tested the anticipated role of augmentation plans in response to regional drought. Seventy percent of the irrigated land in the South Platte system is irrigated with surface water, most of which benefits from senior water rights. Groundwater wells drawing from the South Platte alluvial aquifer are generally junior in priority. In 2006, as South Platte flows shrunk in response to drought, well owners continued pumping even as surface water users had no water. Because of an established hydraulic connection between the river and the alluvial aquifer, it was apparent that the wells were injuring senior surface water rights, but because these wells did not have court-approved augmentation plans, many well owners were forced by the state engineer to cease pumping.[13]

Irrigated acreage within the South Platte River Valley has decreased in recent years and is expected to continue decreasing in the decades ahead in response to water rights being transferred from agricultural to municipal and industrial uses in Denver, its suburbs, and other municipalities. To sustain and support the needs of population growth, municipalities have purchased agricultural water rights for as much as $20,000 per AF of historical consumptive use.[14] If agricultural water rights cannot be purchased by municipalities, they can frequently be leased. Leasing currently accounts for 80 percent of water transactions, by volume, along the river.[15] The South Platte's irrigated acreage may decrease by as much as 267,000 acres by 2050.[16] These agriculture-to-municipal changes in use are similar to those feared by the San Luis Valley agricultural community. The process is much more advanced and common along the Front Range urban corridor.

Arkansas River

The Arkansas River heads in the mountains near the center of the state, flows south, then turns east and exits the mountains near Pueblo before crossing the prairie to Kansas. Just upriver from Pueblo, the Arkansas is impounded by Pueblo Dam. The reservoir has a capacity of nearly 360,000 AF, or roughly 40 percent of the river's average annual flow. The reservoir serves irrigation and municipal and industrial needs, as well as provides protection against flooding.

The first documented efforts to farm along the Arkansas River occurred in 1839, when American and Mexican trappers irrigated crops several miles above Fort Bent.[17] Sustained Euro-American settlement, farming, and irrigation along the Arkansas did not begin until the discovery of gold near Denver in 1858. The first canal to divert water from the Arkansas was excavated near Rocky Ford in 1870.[18] Most irrigation along the Arkansas began during the 1870s near Pueblo.[19] The 1880s saw an extraordinary period of irrigation development throughout Colorado and along the Arkansas.[20] T. C. Henry, the same individual who advanced the Rio Grande Canal in the San Luis Valley, and an enthusiastic promoter of irrigation infrastructure throughout Colorado during the era, purchased the Fort Lyon Canal in 1887 and by 1893 had lengthened it to 110 miles. But "reliable summertime flows" on the Arkansas had been fully appropriated by 1883, and by 1887 diversions from the river began to regularly exceed customary flows. T. C. Henry's Fort Lyon Canal subsequently suffered from water rights too junior to provide reliable water along the canal's full length.[21] Today, numerous canals emanating from the Arkansas and extending all the way to the Kansas state line struggle to serve irrigation demands.

Kansas sued Colorado in 1904 for unfairly appropriating the Arkansas River. Colorado's defense, in part, included historical records depicting the Arkansas River as a "losing stream" more than two hundred miles from the mountains. Citing Zebulon Pike and others, and insisting that there was little or no groundwater or perennial tributaries to contribute new water to the Arkansas as it crossed the plains, Colorado argued that Kansas was complaining about natural conditions, not overuse by its state. A separate argument proffered by Colorado was based on its prior-appropriation doctrine. It would be unfair, Colorado argued, to penalize Colorado for having put the Arkansas to beneficial use before Kansas did. For these reasons and others, the U.S. Supreme Court ruled in favor of Colorado.[22]

Kansas was not convinced, however, and returned to the Supreme Court decades later. In a 1943 decision, the court established the Equitable Apportionment Doctrine as a means to allocate the river between Colorado and Kansas. A formal allocation was subsequently negotiated and implemented by the Arkansas River Compact of 1948. The compact mandated that the John Martin Reservoir, completed the following year and located midway between the mountains and the Kansas state line, be used to store river water partially for use by Kansas.

Farmers began using groundwater from the Arkansas River's alluvial aquifer near Pueblo in the 1940s. Sixteen irrigation wells existed along the Arkansas in 1940; thirty years later there were 1,466.[23] Groundwater withdrawals by irrigators resulted in a familiar situation: "Ditch systems with priority rights to surface water dating back to the 1880s were going without water, while wells installed in the 1950s and 1960s were pumping without limitation." The problem festered until the Colorado legislature directed the state engineer in 1965 to begin administering surface and groundwater rights conjunctively. Implementation was slow until the legislature provided additional guidance and authority in 1969. Real change finally came about in 1985 when Kansas challenged the legality of Colorado farmers taking groundwater that Kansas claimed was intended for it under the 1948 compact. A court concluded that nearly all wells installed in the alluvial aquifer in Colorado between 1948 and 1969 violated the compact. Large-scale pumping of these wells now requires that augmentation water be provided.[24]

Groundwater use in the Arkansas Valley is not nearly as advanced as in the San Luis Valley, primarily because the aquifer underlying the latter is much larger than that underlying the Arkansas. The Arkansas is much more like the South Platte than the Rio Grande in the San Luis Valley. Accordingly, as with the South Platte, surface storage dominates in the Arkansas Valley. In addition to the John Martin and Pueblo Reservoirs, more than a dozen additional storage projects have been constructed in the Arkansas basin, collectively capable of storing nearly 1.9 million AF of water, almost entirely for irrigation. The largest storage impoundment is John Martin Reservoir, with a capacity of 618,600 AF. Together, the John Martin and Pueblo Reservoirs have a capacity nearly equal to the basin's next seventeen largest reservoirs.[25] Modern diversions and storage within the Arkansas basin allow about 428,000 acres to be irrigated on a regular basis, accounting for 995,000 AF of consumptive use, or about 2.3 AF/ACRE.[26] At the Kansas state line, the river's average annual flow is just 155,000 AF and varies markedly with operation of the John Martin Dam.[27]

A troublesome series of events regarding Colorado water began in the 1950s when Front Range cities began purchasing irrigation water rights within the Arkansas watershed. The process gained momentum in 1968 when private investors acquired water rights for Twin Lakes Reservoir above Pueblo. In addition to stored water in Twin Lakes, the Twin Lakes company's shares included rights to 504 CFS from a senior transmountain diversion originating west of the Continental Divide. By 1970, 55 percent of the Twin Lakes corporate stock had been acquired and then sold to Pueblo, Aurora, and Colorado Springs for their water rights. By 1980, the cities owned 94 percent of the Twin Lakes shares. Since 1971, another diversion, the Fryingpan-Arkansas Project, has conveyed up to 69,200 AF/YR from the Fryingpan River in Pitkin County and west of the Continental Divide to the headwaters of the Arkansas River. Currently, about half of this water is for agriculture in the Arkansas Valley; the balance is for Colorado Springs and other municipalities.[28]

By 1991, agricultural water rights sales to municipalities resulted in the dry-up of about 56,000 acres of irrigated land in the Arkansas River Valley, or about 18 percent of the previously irrigated acreage between Pueblo and the Kansas state line.[29] Lawrence J. MacDonnell, a longtime observer of water use in the American West, summed it up this way: "[T]here is simply not enough water in the Arkansas Basin to supply *existing* irrigation demands *reliably*"[30] (emphasis added). This is because "the lower Arkansas is a chronically overappropriated river."[31] This condition exists in spite of transbasin diversions that have, similar to the South Platte, added about 129,000 AF/YR of water to the Arkansas drainage,[32] although the rights to much of it have been acquired by Front Range cities.

The Arkansas and South Platte Rivers will experience increased municipal demand as Colorado's Front Range population continues to grow. Between thirty-five thousand and seventy-three thousand acres may be withdrawn from irrigation by 2050, as their associated water rights are transferred to municipal and industrial uses.[33] "Arkansas River Compact requirements and existing uses and water rights result in little to no water availability for new uses."[34] Without new sources of water, change of use represents the only significant sources available to support population growth without major new transbasin diversions. This history of change of use foretells the threats to agricultural communities in the San Luis Valley.

Intermountain West

Ten western states make up what is labelled herein as the Intermountain West.[35] The states vary slightly in their irrigation practices, agricultural histories, and economies, yet their use of water substantiates a principle thesis of this book: agriculture is the overwhelming user and consumer of water in the West. This could not occur unless agriculture had acquired substantial control over the resource through water laws based on the Doctrine of Prior Appropriation. An overview of some distinctive similarities and differences throughout the Intermountain West, and how they compare with Colorado, is appropriate.

It was not a straight line between California's and Colorado's gold camps, yet as prospectors and miners spread through the West farmers followed, settling near mining districts where they had ready markets for their produce and livestock. In every case, in every mushrooming mining camp and town, people needed to eat. And just as Colorado farmers adapted to the limits of an arid land by adopting irrigation, so too did the farmers who settled the remainder of the West. Eight of the ten Intermountain West states have specifically declared that its natural waters are owned by the public. Wyoming and Montana differ slightly with the caveat that their natural waters belong to the state, not the public at large, although one may reasonably assume that the state serves as agent for the public.

All ten states have adopted prior appropriation as the basis for their water laws. Colorado placed prior appropriation into its 1876 state constitution, whereas the others adopted prior appropriation between achieving territorial status and statehood or, in several cases, following statehood. Practicable means of water rights administration were necessary to prevent chaos in arid farm country, and prior appropriation was the means selected to provide that control. Washington initially incorporated both riparian and prior appropriation into its water rights scheme, but ultimately adopted prior appropriation in 1917, twenty-nine years after statehood. Oregon achieved statehood in 1859, yet required fifty years to adopt a formal water rights system based on prior appropriation. Reasonable conjecture suggests that these Pacific Northwest states lagged in adopting prior appropriation because of the sharply differing climates on either side of the Cascade Range that split these states. Both possess a cool, humid climate along the Pacific Coast and a warmer, drier climate east of the Cascade Range. Riparian dogma initially may have been accepted along the coast where precipitation and water are ubiquitous, while east of the Cascade Range surface water can be scarce. Regardless, today all states in the Intermountain West have well-established water laws based on prior appropriation: "first in time, first in right."

Like Colorado, the other nine states allow only the right to apply natural waters to a beneficial use. Ownership of the water, per se, is not transferred from the state to an appropriator. Wyoming, Oregon, and Washington possess a twist to their water rights regime by providing the state with the authority to deny a water right in the case of public interest, however defined in those states. Colorado does not provide a public interest exception, its constitution firmly proclaiming, "The right to divert the unappropriated waters of any natural stream to beneficial uses shall never be denied."[36]

Water rights elsewhere in the Intermountain West are also generally transferrable or salable. A complicating though common distinction is where a water right is considered "appurtenant" to the land where the water right applies. In non-legal language, this means that the water right is assumed part of the land and is automatically included with any transfer or real estate transaction involving the land. Most states allow appurtenant water rights to be severed from the land in a separate transaction after legal constraints are overcome, the most significant being that other water rights, whether junior or senior, cannot be injured by the transfer.

Consumptive use as a measure of beneficial use is also widely used, although definitions of consumptive use vary. An example is that some states include an allowance for conveyance losses between the point of diversion and the point of use or application. All Intermountain West states allow state regulators to shut down a pumping well if it is shown to injure adjacent surface water rights. Colorado accomplishes this with its policy of conjunctive use: coordinating groundwater rights with surface water rights to maximize efficient use of the resource. Colorado's adherence to conjunctive use stands apart from the other states.

Western states got around to adding in-stream flows to their lists of beneficial uses late in the twentieth century. Colorado has accepted in-stream flow rights "to preserve the natural environment to a reasonable degree," for example, fish and wildlife enhancement, as a beneficial use since 1973. Most other Intermountain West states also consider in-stream flow to be a beneficial use. In-stream flow rights in all states can only be held by a state agency. In Colorado it is the Colorado Water Conservation Board. CWCB has developed and holds in-stream flow rights on more than eighty-five hundred miles of state streams,[37] although a critical distinction is that these in-stream flow rights typically apply to relatively small fractions of a stream's flow and frequently are fairly junior. There are no in-stream flow rights in effect on the mainstems of the Rio Grande, Arkansas, or South Platte Rivers, which were fully appropriated long before in-stream flows were legislated as a beneficial use. Comparable histories have taken place in at least five other western states, where this relatively new beneficial use was added after most major streams and rivers were nearly or fully appropriated. As a result, in-stream flow rights throughout the Intermountain West exist almost exclusively in small to moderate headwaters streams.

An interesting and fruitful situation has developed in Colorado since 2001, whereby a nonprofit organization purchases, leases, shares, or trades water rights and then transfers them to CWCB as in-stream flow rights. The Colorado Water Trust, based in Denver, focuses on senior water rights that can usefully enhance in-stream flows during critical times of the year. The trust has enhanced and restored flows in more than 444 miles of Colorado's streams and rivers. The trust endeavors to create innovative means to finance the transfer and differs from CWCB's efforts to establish original in-stream flow rights by working their way through water court.

Colorado's conjunctive use policy, where groundwater and surface water rights are coordinated in the prior-appropriation system, appears unique among the ten western states. The other nine can prohibit groundwater wells from injuring adjacent surface water rights, but this is not the same as coordinating individual priorities. The absence of conjunctive use policies elsewhere in the Intermountain West results in less efficient management of limited water resources.

Water use practices by agriculture throughout the Intermountain West are similar, so looking into how, and how much, water is annually diverted to irrigated farmland, and how much water is consumptively used, requires statistics. Fortunately, these statistics are available.[38] How much water is diverted for irrigated agriculture varies by priority date, water availability, the number of acres under irrigation, the conveyance loss that must be made up between point of diversion and point of use, the amount of water required by particular crops, the antecedent soil moisture prior to irrigation, weather, and the irrigation method employed. These factors and others affect how much water is diverted both instantaneously and over the course of a growing season. A report by the U.S. Geological Survey relates that seventeen states in the American West, including Colorado, are responsible for 83 percent of total irrigation withdrawals and 74 percent of the irrigated acres in the United States.[39] (The seventeen states include those on the Great Plains; the definition of the Intermountain West used here does not apply.) In the Intermountain West, 43 percent of irrigation diversions are groundwater. Given that groundwater was rarely recognized or used as a significant source of irrigation water by mid-nineteenth-century settlers, this represents phenomenal growth in the use of what in many cases is a nonrenewable source. The aquifers underlying the San Luis Valley and South Platte and Arkansas Rivers are shallow and recharge readily in response to leaky canals and ditches, irrigations, and floods. Deep aquifers, however, generally recharge very slowly, over eons, typically measurable only in geologic time. Extracting water from them is little different than mining. When it has been exhausted, it is, for all practical purposes, gone.

The West's increasing reliance on groundwater for any and all applications is therefore risky, because there is no apparent substitute when nonrenewable groundwaters have been consumed. Even so, several western states are proceeding in that direction, including Nevada, New Mexico, and especially Arizona. Efforts by Las Vegas to acquire groundwater rights in northern Nevada's lightly populated valleys are well documented, although for now the city is relying primarily on the Colorado River. Farmers on New Mexico's eastern plains continue to draw down deep aquifers without an alternative for when they hear the figurative slurping sound of a dry well. Arizona closely monitors and controls groundwater withdrawals near its population centers of Phoenix and Tucson, and the state has had some success in limiting groundwater declines. Yet there are almost no groundwater use regulations in place elsewhere in the state other than an elusive "reasonable use" doctrine.[40] Corporate farms have located in lightly populated valleys and drilled deep wells "to irrigate tens of thousands of acres of hay, corn, pistachios and other thirsty crops."[41] The long-term ramifications of these actions are unknown but likely problematic.

The national average diversion rate in acre-feet (AF) per acre is 2.07, that is, just over two feet of water is diverted each year for each acre of irrigated farmland. It is understandably lower in the humid, rain-rich East—for example, 0.77 AF/ACRE in Maryland—where natural precipitation is ample. In the Intermountain West, the average is 3.68 AF/ACRE, 78 percent higher than the national average, and ranges from 2.24 AF/ACRE in Washington to 5.16 AF/ACRE in Arizona. The latter has the highest value of any state in the nation.

States primarily employing less efficient irrigation methods (e.g., flood or furrow) understandably have the highest application rates, because less efficient irrigation necessarily requires the diversion and application of greater amounts of water to cultivate a given crop. Arizona's application rate of 5.16 AF/ACRE seems to be higher than other western states because only 20 percent of its irrigated farmland is efficiently watered by sprinklers. Another reason Arizona diverts so much water per irrigated acre is because the state's climate allows at least two crops each year. Wyoming is similar in that only 17 percent of its irrigated cropland is watered with sprinklers; its diversion rate is 4.53 AF/ACRE, third highest in the nation. Conversely, Washington farmers irrigate 80 percent of their cropland by sprinkler and divert only 2.24 AF/ACRE of water. The general relationship is unambiguous: more efficient irrigation methods require the diversion of less water. This relationship does not alter the fact that more efficient irrigation methods, like sprinklers, increase consumptive use while simultaneously reducing return flow. Similarly, the relationship ignores the fact that more efficient irrigation methods require greater capital investment by farmers. Not every farmer can afford to install and operate more efficient irrigation systems, even when federal financial assistance may be available.

Stark numbers are likewise apparent if the number of irrigated acres for each major type of irrigation method—sprinkler on one hand, surface or flood on the other—are viewed using application rates and assumed irrigation efficiencies.[42] Analysis yields an estimate of annual consumptive use of water by irrigated agriculture in the West and obviously varies due to the number of irrigated acres in each state. Total annual consumptive use ranges from a low of about 1.143 million AF/YR in Nevada to as much as 6.96 million AF/YR in Colorado. Cumulatively, the ten Intermountain West states may be responsible for more than 40 million AF of consumptive use per year, an amount that is roughly twice the annual flow of the Columbia River. Compared to the 62 million AF diverted for irrigation each year in these ten states,[43] it is apparent that irrigated agriculture is responsible for diverting a huge proportion of the region's water. A recent research effort by numerous universities and researchers also concluded as much: "Irrigated agriculture clearly has a dominant influence on river flow depletion across the western US.... More specifically, irrigation of cattle-feed crops (including alfalfa and grass hay and haylage, corn silage and sorghum silage) is the single largest consumptive user at both regional and national scales, accounting for 23% of all water consumption nationally, 32% in the western US and 55% in the Colorado River Basin."[44] Unambiguously, irrigated agriculture is responsible for an astounding amount of consumptive water use in the West.

The difference between the quantity of water diverted from the West's rivers and streams and the quantity consumptively used is an indicator of inefficiency that in many—if not most—cases such numbers can be reduced. This same point was made by Lawrence Mac-Donnell in his influential work *From Reclamation to Sustainability:* "Making more efficient use of water, measured in terms of reducing the gap between the amount of water diverted and the amount usefully consumed, is not an end in itself. It is a means, potentially, of achieving other desired objectives."[45] One would not have to go far to identify other attractive and economically rewarding uses for the Intermountain West's streams and rivers.

CHAPTER FOURTEEN

A New Normal

Water is growing increasingly scarce, yet we continue to act as if it will always be abundant.

— Edward B. Barbier, *The Water Paradox* (2019)

Growing food crops is impossible without water. Water is irreplaceable, its value undisputable. And where nature fails to deliver sufficient quantities in a timely manner, it has always been obtained by bringing water to the fields. Colorado's 1876 constitution guarantees the right to divert water from the state's phenomenal streams and rivers while simultaneously declaring that these waters belong to the public. The first and greatest beneficiary of this seemingly odd juxtaposition has always been agriculture. In the 1850s, for decades to follow, and to this very day, the set price for water in the San Luis Valley and throughout the West has been near zero, the only real cost being for diversion and conveyance infrastructure. There has always been a gaping divergence between water's cost and its value.

Edward B. Barbier, professor of resource economics at Colorado State University, has warned, "We ignore the signs of growing water scarcity until sudden and unexpected shortages force us to take drastic measures to curtail excessive use."[1] Fresh water has always been limited in supply, especially in the West. This characteristic has never been secret.

Barbier also recognizes another fact of life that summarizes irrigation in the San Luis Valley: "the prices which most users pay for water reflect, at best, its physical supply cost and not its scarcity value. Users pay for the capital and operating costs of the water supply infrastructure but, in the USA and many other countries, there is no charge for the water per se. Water is owned by the state and the right to use it is given away for free."[2] Barbier's assertion accurately describes water's development and exploitation in the San Luis Valley, the rest of Colorado, and the arid American West. The construction and operation of most of the valley's reservoirs and irrigation canals, laterals, and ditches—the infrastructure—have been borne largely by the agricultural community, but there has never been a cost for the water on behalf of the state's citizens. Barbier qualifies his description by asserting that water is cheap either because the infrastructure is inexpensive, or because the water is subsidized.[3] A cheaper subsidy than free is difficult to imagine, but it has been the status quo in Colorado and the American West since the first settlers arrived in the middle of the nineteenth century.

Nonetheless, things began to change over the past decade in the San Luis Valley. Water's scarcity, due to overdevelopment, drought, or both, forced surface water users to realize that there was an inadequate supply to satisfy everyone's needs, especially when groundwater users jumped ahead of more senior water rights holders. It has been a trying period for all, eventually forcing an accommodation upon the irrigation community, where groundwater interlopers were forced to provide augmentation water to senior rights holders through the valley's innovative groundwater subdistrict system. The medium was money. Subdistrict members now pay for the water they previously used for free.

Agriculture has controlled the Rio Grande in the San Luis Valley since before statehood. The river has been measured, engineered, and meted out to irrigation for so long that it seems natural to the valley's residents. But after accrued water rights have consumed a river's flow, there is nothing left for other uses. Colorado labels stock watering, domestic and municipal consumption, manufacturing, fish and wildlife, snowmaking, truck washing, and other practices as beneficial uses under state law.[4] Since 1973, in-stream flows have also qualified. The point is not that uses other than agriculture are prohibited or discouraged; it is that agriculture was first in line and has from the beginning diverted and consumed the greatest amount of Colorado's, and the West's, water. In the San Luis Valley, there is no doubt that agriculture controls nearly all water. Colorado and the West mimic the valley in controlling and managing water.

During late summer, when irrigation demands are greatest, some of Colorado's streams and rivers, if they flow at all, are single-purpose waterways, an occurrence that makes apparent a significant corollary to unlimited diversions from a stream or river, a corollary that points to the role played by the State Engineer's Office. The office does more than parcel out water rights. It also manages the state's streams and rivers as conveyance channels, moving water from one irrigation diversion to the next. Some Colorado streams and rivers serve no purpose other than to transport irrigation water. It is the same throughout the West.

Can Colorado and the West accommodate agriculture's continuing domination of the region's water resources? Should agriculture have first dibs on all water, all the time, in all places? The frontier conditions that fostered agriculture's place at the head of the line no longer exist. Any moral imperative claimed by agriculture's loudest proponents no longer carries the argument. It is equally apparent that people need to eat, and farmers require certainty of water to function and prosper. Farmland without water borders on worthless, while with an adequate and reliable water supply the land can bear fruit and support families and businesses, a relationship that has existed for centuries. Water rights play to a banker's willingness to loan money to a farmer or rancher. Bankers rely on the value of water rights, in part, for the collateral required for these loans. Adequate and reliable water underlies the certainty of hundreds of small communities in the West. Without water, small businesses and towns would blow away with the vanishing farms they once supported. Even so, Colorado may wish to see more water flowing in its streams and rivers. Fishery and recreation interests will grow louder. Colorado sees benefit in healthy river ecosystems and greater natural flows, and there is no reason to believe this attitude does not exist throughout the region. It does not matter for now, however. The economic faction that controls the greatest share of Colorado's water is agriculture. State law benefits agriculture.

Water use over the past century has also resulted in more environmental change in Colorado and the West than anyone might have predicted in the nineteenth century—if making a prediction then was thought worthwhile. Compare the San Luis Valley's bogs, meadows, and meandering sloughs of 1806 with the hundreds of circular irrigated fields of the twenty-first century. In the valley's perpetual pursuit of water, fully 50 percent of the valley's wetlands have disappeared since the 1980s.[5] While our collective memory of how watersheds appeared and operated even a generation ago is a function of incremental environmental change, it reflects the tradeoff we have been willing to make for homes, farms, and ranches.

Maybe the religious scholars and philosophers of past millennia were wrong and the earth is not for humankind's dominion. Maybe centuries of religious and philosophical exposition were self-justification on a colossal scale, and humans expand our ecological niche like any other organism. Maybe the gods are curious and want to see what we do with it. But maybes do not matter. Irrigators in the San Luis Valley must deal with reality. The principles of irrigation, consumptive use, and return flow apply. What makes the San Luis Valley differ from Colorado's other major agricultural centers is the complexity of its manmade aquifer on which the valley's economy rests. Everything else, everywhere else, is analogous. If much of the West is water-short, Colorado's San Luis Valley is inordinately blessed. The Rio Grande has been, on whole, generous.

◆ ◆ ◆

In considerably less than two centuries, at least four events demonstrated where and how irrigation and water use have exploited nature's largesse and societal harmony in the San Luis Valley. Since the valley's first settlers arrived in the 1850s, the valley's streams and river have been repeatedly overtaxed. In each instance, a physical or legal limit was reached.

The first event began in the 1880s when entrepreneurs constructed extensive canal systems to spread water across the valley floor. Within two decades, in response to the onset of drought, farmers dewatered the river, leaving little for New Mexico Territory, Texas, and the Republic of Mexico. Federal intervention was required to achieve a fair apportioning of the river. Even that took another forty years.

A second event soon followed. By the early twentieth century, the valley's unlined canals and ditches allowed the water table to rise nearly to the ground surface in the Closed Basin. Subirrigation was recognized and came into practice and flourished. But irrigators overlooked the natural salinity of the soils, rising groundwater brought these salts to the surface, and by 1915 farms with damaged soils had to be abandoned. Subirrigation nevertheless continued in portions of the valley and remained a principal means of crop watering into the 1950s.[6]

Beginning in 1951, in the jaws of yet another drought, a third threshold was achieved when Colorado began to under-deliver its Rio Grande Compact allocation. Colorado continued to shortchange downriver users until New Mexico and Texas threatened Colorado with legal action before the U.S. Supreme Court in 1968. Colorado eventually relented as part of a stay of litigation and restarted its full delivery obligation.[7] Notice had been given that the compact was real and enforceable. Today, irrigators, regulators, environmentalists, and recreationalists concede that, given the opportunity, the valley's irrigators would again regularly drain the Rio Grande, at least during the irrigation season. Only the compact keeps the river flowing to downriver users.

A fourth limit was initiated in the mid-twentieth century when valley irrigators realized that seven thousand miles of leaky canals and ditches had created a tremendous aquifer beneath their farms that could be used more efficiently than traditional surface reservoirs in the neighboring mountains, especially when used in association with sprinkler systems. More efficient than either flood irrigation or sub-irrigation, sprinklers led to more efficient water use and increased crop yields. As irrigators increased their reliance upon groundwater, however, junior groundwater rights overwhelmed senior surface water rights. State regulators did not have the tools to administer conjunctive groundwater and surface water rights in the valley. Tensions increased between surface and groundwater users; they worsened after 2002 when the valley entered another of its infamous and recurring droughts. A dwindling Rio Grande left farmers with senior surface water rights dependent on paltry surface flows, while groundwater users with junior rights continued to exploit a shrinking aquifer. Only innovative thinking attained a limited degree of equanimity, and even that required the irrigation community to begin charging itself real money for out-of-priority water use. The real economic value of water finally began to reflect its scarcity.

One may reasonably conclude from these four events that, when water is free, agriculture will exploit it until demand outstrips supply. Alternatively, if drought limits the supply of water, agricultural demand will increase water's value. In both cases, an increase in the cost, or economic value, of water becomes apparent.

◆ ◆ ◆

The future of irrigated agriculture in Colorado, writ large, promises continuing change. Subtle challenges will sneak up in the years and decades ahead, but major challenges are already obvious. For one, water rights administration throughout Colorado has grown increasingly complex, evidenced by the recent creation of groundwater subdistricts in the San Luis Valley and the need for a computer model of the valley's hydrology. Both help the division engineer administer water rights and allow irrigators to use surface and groundwater conjunctively. The days when water rights can be fairly and accurately administered from horseback have passed.

Another challenge facing Colorado irrigators is ongoing improvements in technology. Federal cost-share programs encourage such improvements,[8] even though sprinkler use in the San Luis Valley has muddied things by reducing return flows to the valley's primary aquifer. The matter is rendered more complicated by the fact that less than half—possibly far less than half—of groundwater decrees in the San Luis Valley have a limit on annual groundwater withdrawal.[9] Pumping *rates* are specified, but not allowable annual quantities. Consumptive use has increased over time, but no one knows how much because almost no one attempts to measure it. Division engineers are directed to manage water rights "to the decree"[10] under the implicit and reasonable assumption that there is a practical upper limit to how much water any field and crop can consume. Overwatering can limit crop yields just as under-watering does. Add longer growing seasons and improved crop genetics to the mix, and one should expect return flows to underlying aquifers and adjacent streams and rivers to decrease. The impact is easily overlooked in the San Luis Valley because of its size and the huge volume of water in the heavily used aquifer, but elsewhere in Colorado and the West, in smaller hydrological systems, impacts from reduced return flows will manifest more quickly as dry streambeds. What can be concluded from this cascade of nature and human practice is that irrigated agriculture's continued water diversion practices will result in a fluctuating crapshoot for local environments. A worthwhile change would be to bring allowable diversions more in line with consumptive uses.

Another predicament facing Colorado and the West is the need and cost to rehabilitate hundreds of dams and reservoirs. Just in Colorado they number over nineteen hundred,[11] many—perhaps most—owned and operated by farmers and ranchers, and those sites will fall under some limitation as they near the end of their engineering lives. Rehabilitation and safety improvements will not be readily or easily afforded by their owners. Financing the needed rehabilitation of these dams and reservoirs was frequently available from state and federal government sources in the past, but that is less likely in the future. Help is essential. Assistance *may* be available to private owners through selected agencies and interest groups keen on seeing more water in the rivers, and who can weigh in with political support to help pry dollars from penurious state and federal pockets. They may even have their own capital to contribute. Collaboration and cooperation can help save ailing infra-structure, provided owners are willing to modify water use practices and voluntarily enter into partnerships to provide multiple benefits. The model for this is the Rio Grande Cooperative Project, embraced by the San Luis Valley Irrigation District and Colorado Department of Parks and Wildlife. A slow process, improvements to reduce seepage through the dam were completed in 2013 and funded with grants from the Colorado legislature. Necessary safety enhancements to the outlet structure were funded *in part* by the legislature, but also required the irrigation district to contribute toward funding the project. The outlet structure was completed in late 2019.[12]

Against these multiple challenges, the element that bodes well for the San Luis Valley and the future of its irrigation economy is its inher-ent sense of community and learned willingness to collaborate. This attitude grew in reaction to the AWDI episode in the 1980s, but it seems likely that the valley's geographic limits and surrounding mountains engender a certain cordiality when under duress. Multiple generations, living and working together for more than a century, have learned how to get along in a limiting desert environment. Certainly there are bouts of frustration and anger, but a western counterpart to the historic Hatfield-McCoy feud does not exist in the San Luis Valley.

More complex water rights administration, the demands created by advancing irrigation technology, and the need to rehabilitate aging and worn infrastructure will increase real costs to agriculture. One can rail against the march of civilization and yearn for the good ol' days, but the simple explanation is that, directly or indirectly, the cost of water to agriculture is beginning to reflect its true value. It was too easy in the past to misuse or overuse free water. Members of the irrigation com-munity who can "pencil out" profitable and rational operating plans in this environment will have a distinct advantage.

At the economic continuum's other end are marginal producers, many of whom hold jobs away from the farm or ranch so they can afford to continue farming and ranching. The U.S. Department of Agri-culture projected in 2018 that, "On average, 82% of U.S. farm house-hold income is expected to come from off-farm work...up from 53% in 1960."[13] Whether of necessity, stubbornness, a passion for agriculture, or a combination of these factors, members of this group are making a lifestyle choice where agriculture provides only a portion of home incomes. That lifestyle choice has real consequences in terms of water use in Colorado and elsewhere.

Farmers and ranchers throughout the West, not just those in Colorado, also are aging. The average age of contributors to the agricultural economy is rising steadily. Participants are having difficulty interesting the next generation in taking over the family farm or ranch. While this phenomenon is possibly not widespread in the San Luis Valley, where even young farmers may prosper, facilitators of suburban and urban population growth will eagerly purchase water rights from aging farmers and ranchers who feel they have few alternatives as financially attractive as selling.

The greatest current challenge to agricultural water use in the American West is probably population growth and accompanying demands for domestic and municipal water. AWDI's defeat in the 1980s did not terminate all efforts to export water from the San Luis Valley. The AWDI story was merely one chapter in a book of unknown length. It is a process without an imminent end. In Colorado, at least 80 percent of the state's natural supply of water is diverted and consumed by agriculture.[14] It may be well over 95 percent in the San Luis Valley. As America continues to move west, a trend that will not change anytime soon, people will require water to support communities and businesses as they settle, congregate, and urbanize, especially along Colorado's Front Range. When they turn the faucet, they expect drinkable water to flow. Taps will be turned; headgates will be opened; water will be expected; surprises will not be tolerated or rewarded. Indeed, as Barbier has concluded, "The only way to meet the growing municipal, industrial, recreational and environmental water demands in the western United States will be through reallocating water out of agriculture."[15] The politics of water in Colorado and the West has always been a two-fisted brawl. It is only going to get rougher. Water in the American West is finally being recognized for its true economic value. And the speed at which this is occurring increases daily.

Twenty minutes before the scheduled commencement, the meeting room was nearly full. The regular monthly meeting of the Rio Grande Basin Roundtable was taking shape. Men and women gathered in small groups, conversing about the weather. Half of the attendees appeared to be under forty years of age. It was January 2018. Outside it was sunny and fifty degrees, not the customary zero with snow on the ground. Down-filled Carhartt jackets and felt-lined Sorel boots had been left at home. Individuals wore denim jackets or no jackets at all. Some just wore T-shirts.

Nathan Coombs, an affable rancher and roundtable chairman, called the meeting to order. His predecessor, Mike Gibson, had retired. Introductions ensued, followed by a quorum call to ensure the meeting could take official actions. A quorum being present, Coombs announced that the annual election of officers was the first item on the agenda and handed control of the meeting to a non-officer. The current officers agreed to serve another one-year term, no new nominations were offered, and the meeting unanimously reelected Coombs and his cohorts. The election took five minutes. After laughter and several good-natured insults, Coombs retook control of the meeting.

First to report was Cleave Simpson, who, following Steve Vandiver's retirement, had recently been appointed general manager of the Rio Grande Water Conservation District. Simpson is burly, like Vandiver, but without the moustache. He is serious yet smiles readily. He was raised in the valley, acquired an engineering degree at Colorado School of Mines, and spent twenty years working in the lignite fields on the Texas coast before returning to the valley. In addition to his position with the district, he operates several farms with his father, wife, and son. He looks forward to his son becoming the fifth generation of Simpson farmers in the San Luis Valley.

Simpson announced that formation of Groundwater Subdistrict No. 5 had been approved by the water court the month before, drawing more than three hundred new well owners into the valley's groundwater subdistrict system. Like the other four subdistricts, the new one is committed to rigorously managing the groundwater resource that irrigation depends on. More than 96 percent of affected well owners in the valley had seen the benefit of joining subdistricts.[1] The new subdistrict members would now pay to pump and use groundwater out of priority. All hoped to achieve a degree of success like that attained by Subdistrict No. 1, which, since its formation six years earlier, had seen groundwater pumping decline by one-third. Incurring new fees had become an effective incentive for valley farmers to use the resource judiciously. It cost more to farm, but the alternatives were costlier and could put them out of business.

The roundtable meeting continued. Disappointment rose when it was revealed that statewide roundtable funding for 2019 and beyond would be constrained because of a recent Colorado Supreme Court ruling. The state's nine roundtables were assured that the current year's funding was certain, plus the Rio Grande Basin Roundtable had its own reserves to draw upon, so the meeting agreed to go forward with a vote on financing proposed projects. Roundtable members listened, politely asked questions, then unanimously approved all five requests. More than $1.3 million was committed, a healthy portion from the roundtable's own account.

The final speaker was Craig Cotten, Division 3 Engineer. His monthly updates on flows in the Rio Grande and its tributaries, snowpack, and the status of obligations to the Rio Grande Compact are always meeting high points. His last update in November had been optimistic about the Rio Grande's final 2017 flow: 690,500 AF, or 108 percent of the long-term average. In fact, flows had been near or above average for the preceding three years, news that was welcomed at the time and helped explain why the valley's primary aquifer continued to recover. Still, the room held its collective breath. The winter of 2017–2018 thus far had been mild. Their collective prayer was that Cotten would report something more positive than what they had been reading on the internet.

Cotten's prediction for the 2018 growing season was as gloomy as any in the room feared. Snowpack in the Rio Grande's headwaters was 31 percent of normal, less than in 2002 and 1977, the two driest years in recent history. Unless winter storms began dumping more snow in the mountains, Cotten forecasted Rio Grande runoff for 2018 would be about 345,000 AF, little more than half the long-term average,[2] and that the reaches of the Rio Grande between Del Norte and Alamosa, where the valley's major canals had their headgates, would be dry by midsummer.

A wave of groans rose in the room. Many mentally revisited 2002 and the beginning of the ongoing, devastating drought, when the Rio Grande yielded less than 164,000 AF, the lowest in recorded history. They had been hoping that, somehow, 2017 had been nature's assurance that wetter years were ahead. However, in spite of recent upticks in annual runoff, the trend since 2001 had been downward, averaging 15 percent less than the pre-2001 average. At least one outside study had already concluded that, in response to a changing climate, flows in the upper Rio Grande watershed could decline by 25 percent by 2100.[3] There was no doubt that 2018 was looking grim.

The meeting ended and attendees began to wander out into a chill winter afternoon. Lingering conversation spun on Cotten's gloomy runoff prediction. Cleave Simpson took his time leaving. He chatted with loitering friends but kept his worries to himself. During the approaching 2018 growing season, renewed drought would trigger more groundwater pumping in the valley. Would the groundwater subdistricts be effective? Would the aquifer's recovered groundwater storage be lost again? He feared that in one very dry summer everything regained since 2012 could disappear.

◆ ◆ ◆

Simpson's fears were eventually realized. The effects of the extremely low snowpack from the 2017–2018 winter were worsened by an exceptionally dry irrigation season. Total river flow during 2018 was just 44 percent of its long-term average, the fourth worst runoff year on record. Having little surface water available during the growing season, irrigators in the valley used groundwater. Groundwater stored in the valley's unconfined aquifer beneath Subdistrict No. 1 declined by 260,000 AF during the year, or nearly all that had been gained since beginning operation in 2012. It was one of the steepest single-season declines in more than forty years.[4] By the end of 2018, the aquifer was about 600,000 AF *below* what the state engineer had mandated must be recovered by 2032 and what the district had agreed to in its court-approved operating plan.[5] Another indicator of the drought's severity in 2018 was that the Closed Basin's contribution to Colorado's Rio

Grande Compact obligation decreased to only 8,000 AF for the year, less than 10 percent of what was originally predicted.[6]

As the valley entered the winter of 2018–2019, another ill-fitting shoe dropped. A new proposal surfaced to annually export 22,000 AF of groundwater to the suburbs south of Denver. The proposal differed from the AWDI plan that had been successfully fought in the 1980s. The AWDI plan was based on perfecting a new water right in order to export 200,000 AF per year. The new plan, posed by Renewable Water Resources (RWR), is potentially easier to execute. The company's backers include a former Colorado governor and state legislator and at least one farmer from the north end of the San Luis Valley. The plan is predicated on the purchase of a portfolio of surface and groundwater rights around the valley, and with a series of water rights retirements, reductions in the number of acres irrigated, and related swaps to free up 22,000 AF of water to export by pipeline. Proposed water rights retirements would allow some water to be retained in the valley and provide a counter to ongoing water shortages. The valley's pervasive poverty would likewise see improvements, RWR's backers argued. The San Luis Valley has some of the poorest counties in Colorado. Between 21 and 30 percent of its residents live below the poverty line, 10 percent higher than the state average.[7] Surely the money inflow would contribute a much needed economic boost to the valley and its less affluent occupants.

RWR estimates that water rights in the valley can be purchased for about $2,000 per AF, suggesting an initial capital investment of $44,000,000 just for rights to the water. Considering a pipeline cost of $550 to $600 million and ten years to complete the project, it is obvious that RWR is engaging in an extremely ambitious business proposition. Water in the Denver area can garner between $10,000 and $30,000 an AF, or between five and fifteen times what water rights might sell for in the valley. A significant element of the plan is obtaining and manipulating decreed senior water rights, indicating that the issue will hinge in part on private property rights to the water and whether the owners of those rights can and will sell them. RWR has assured the public that more than forty of their neighbors are, indeed, interested in selling their water rights. In the rural West, private property rights are sacrosanct, and interference with an individual's right to do what he or she wishes with their private property tends to be quite unpopular. As long as the Rio Grande and San Luis Valley have water, valley farmers and ranchers know they will be targeted by developers. Also apparent is that at least a few of their neighbors are interested in selling. In spite of these caveats and assurances, in January 2019 the Rio Grande Water Conservation District board refused to support RWR's plan.

RWR is not giving up and continues to talk with valley farmers about possible improvements to its plan, including raising the per-AF purchase price and the likelihood of future royalty payments. RWR also continues to approach Front Range water users about their interest in future water purchases. The Front Range is estimated to need another 300,000 AF of water by 2050 to keep up with population growth.[8]

◆　◆　◆

Nature is nothing if not changeable. As the valley absorbed news of RWR's water export proposal, one of southwest Colorado's longest and most severe droughts began to lighten. A stream of winter storms began to course through the state and the Rio Grande's headwaters. Originating from the Pacific Ocean, the storms were laden with moisture. Snow fell and kept falling. Ski areas began to enjoy a banner year, county and state snowplow drivers did double duty, and highways over several of the state's highest mountain passes were closed for weeks at a time because of avalanches. By late February 2019, optimism in the valley was obvious and rising. More snow in the mountains meant more water for the growing season. Seven groundwater subdistricts either existed or were expected to be in operation by 2020, incorporating between 95 and 98 percent of the irrigated land in the San Luis Valley.[9]

In the face of this optimism, Cleave Simpson nonetheless suggested that between a hundred thousand and a hundred fifty thousand acres in the valley could no longer be supported by the highly variable water supply and strict aquifer recovery requirements.[10] But news about the upcoming season remained good. By mid-March, the Division 3 Engineer had assessed the snowpack in the surrounding mountains, compared his division's calculations with those from sister agencies, and predicted that total runoff from the Rio Grande in 2019 would be about 760,000 AF, the largest since 2007.[11] Maybe, some opined, the drought was over. Or maybe not. Who knew for sure? Even in the face of local optimism, Simpson was worried. Water seemed to be plentiful—this year.

By August, groundwater pumping had been so sufficiently brisk that any hopes Simpson had had for the unconfined aquifer's recovery had been limited to just 150,000 AF, a lesser amount than it had declined the year before.[12] For the total 2019 year, the aquifer eventually recovered about 180,000 AF.[13] The irrigation community evidently had chosen production over the looming requirement to prove that the aquifer could be managed sustainably. The farming community was content and the local economy was humming.

Nature swapped directions again and, by the time the irrigation season began in April 2020, all indications pointed to the drought's wrathful return. There was no constancy in the Rio Grande's flows or lessening of the drought's threats to the local economy. The valley felt whiplashed. Water levels in the unconfined aquifer began dropping in January, well before irrigation season began. Simpson's job requires him to take the long view, but nature seemed to be toying with the Rio Grande and San Luis Valley. Nothing would be simple or easy.

Nature's challenges are rarely announced in advance, especially when water is concerned. Variability and unpredictability are givens. Droughts have occurred in the past, and there is every reason to believe they will occur in the future. Indeed, they may worsen in response to predicted climate change. This all must be dealt with while fighting off what many in the valley brand as water bandits from the Front Range. Meanwhile, the irrigation season begins in April, headgates open, canals and ditches fill with water, and the San Luis Valley's growing season starts anew. In Colorado and the American West, the struggles between aridity and a human insistence on living in a dry land promise to remain a gamble.

ACKNOWLEDGMENTS

This project began with observations and conversations with neighbors and friends when I was farming in a western Colorado valley, where irrigation is the lifeblood of the local agricultural economy. The fact that considerable water was being diverted from the local river for irrigation was no secret, but I became curious about how and why the river dried up late every summer and whether it had to be that way. From that curiosity grew this volume. I have concluded my research and writing. As interview transcripts, references, maps, notes, and pieces of manuscript gradually overwhelmed my desk and poured onto the floor, I realize how dependent I was on the generosity of others. No one ever refused my plea for information or assistance. A willingness to put aside one's immediate tasks to help another is a trait shared by many in the West. It is time to acknowledge those individuals.

Three persons especially assisted and advised me over the years. In fact, two of them retired while I was plodding my stubborn path. Travis Smith, former superintendent of the San Luis Valley Irrigation District, spent many hours describing irrigation in the San Luis Valley and how water rights are administered on the Rio Grande, filling in gaps in my research and reading, and showing me around the district. Twice, we made the lengthy drive to the Rio Grande Reservoir deep in the San Juan Mountains. Steve Vandiver, former Division 3 Engineer, then general manager of the Rio Grande Water Conservation District, generously and candidly explained irrigation's intricacies, as well as upgraded my understanding of water rights administration in Colorado. I met no one possessing greater expertise about water rights and irrigation in the valley and their history over the past forty years. Cleave Simpson, Vandiver's successor at the Rio Grande Water Conservation District, picked up where Vandiver left off and kept me updated on the status of the valley's aquifers and the slow, methodical progress of the groundwater subdistricts as they came online. If I have made any misstatement of facts or gross errors of interpretation in these pages, it certainly is my fault and not theirs.

Rio de la Vista provided useful background information at the beginning and toward the end of my efforts. Currently head of the Salazar Rio Grande del Norte Center at Adams State University in Alamosa, few persons bring greater passion to their work and a deeper desire to keep San Luis Valley residents informed about the status of the Rio Grande and its wetlands.

Not all the experts consulted reside and work in the San Luis Valley. In Denver, Mike Sullivan, Deputy State Engineer with the Colorado Division of Water Resources, graciously answered my questions and explained practical issues surrounding irrigation in the San Luis Valley and administration of Colorado water law. His knowledge of problems as he observed and dealt with them at an administrative level ironed out some misconceptions I might have had. He also offered valuable facts and statistics. Another expert who assisted me in my efforts to understand the surface and groundwater systems in the San Luis Valley was Eric Harmon, a principal with HRS Water Consultants and one of Colorado's premier groundwater engineers. He has worked in the valley over decades in pursuit of a better understanding of its water; there may be no groundwater professional who possesses a better grasp of the valley's water resources.

David Robbins, attorney with Hill & Robbins, P.C., one of Colorado's most experienced water lawyers, has represented the Rio Grande Water Conservation District for years. He amiably and graciously sat with me for several hours as I quizzed him on issues surrounding water law and irrigation in the San Luis Valley and on the Rio Grande. Also in Denver, David Nickum and Drew Peternell, with Trout Unlimited, willingly spent several hours explaining TU's aspirations for Colorado's streams and rivers and their cold-water fisheries.

Gregory J. Hobbs, former Colorado Supreme Court Justice and currently Distinguished Jurist in Residence at Sturm College of Law at the University of Denver, offered encouragement, advice, and key pieces of information several times when we crossed paths at conferences. He particularly improved my understanding of the legal distinction between beneficial use and consumptive use. If I have nonetheless warped his lesson in these pages, it is my fault, not his.

Dr. Willem Schreuder, the brain behind the consulting firm Principia Mathematica, several times affably discussed and explained his computer modeling of the San Luis Valley and how it is being used to guide the valley's groundwater subdistricts in their pursuit of effective conjunctive use. I am among many who believe his efforts will lead to more efficient use of the upper Rio Grande.

Other individuals living and working in the San Luis Valley or on Colorado's western slope provided timely advice and perspective, including Cary Denison, Mike Gibson, Heather Dutton, Mike Blenden, John Alves, Eugene Vasquez, Andrew Valdez, Kevin Terry, Ralph Curtis, Mike Wellman, David Hayden, and Paul Robertson. Thanks to all for your generosity of spirit and time and your willingness to share your knowledge. As well, I am grateful for help in locating needed reference materials from staff at the Denver Public Library and John F. Reed Library at Fort Lewis College in Durango, Colorado. They saved me hours of unproductive wandering among the stacks.

William Crank, a great friend and clear-sighted reader, reviewed and commented on numerous drafts of the manuscript as it took shape. A sharper editorial eye would be difficult to find. I still owe him for three days of backbreaking help in moving gravel-filled irrigation pipe. Gene Reetz, a knowledgeable hydrologist and good friend, provided welcome comments and advice on a key chapter toward the end of the writing process. And Jon Harvey, assistant professor of geology at Fort Lewis College, ably produced the maps included in these pages. He was always willing to tweak them as the manuscript took shape. I cannot imagine a book of this type without competent map work, and Jon made it happen.

Finally, I gratefully acknowledge the unending assistance of Jean Aaro, my wife and beloved editor. She read the complete manuscript many times during its evolution, encouraging me always to simplify and focus on clarity. Her value in my life goes far beyond editorial advice.

I am not an attorney and cannot and will not claim that the legal definitions, judgments, and opinions concerning Colorado water law as espoused in these pages are without error. But I have been exposed to numerous water users and irrigators over the years and have participated in many discussions on related topics, sometimes with attorneys, many times without. I freely opine that many grumpy farmers, having irrigated for decades, may have as much working knowledge of Colorado water law as anyone in the courtrooms. Argue with a shovel-wielding farmer wearing irrigators' boots at your peril.

To those whom I inadvertently neglected to mention, I offer my sincere apology. I hope you remember and know who you are and accept my gratitude even if I fail to acknowledge you individually. I was the grateful recipient of much help, advice, and information along the way. The simple fact remains that while I believe these pages to reflect truth and candor, they nonetheless constitute my version and interpretation. I am human, with flaws. And where I offer an opinion in these pages that may run contrary to those who helped me along the way, I trust the distinction is obvious. I will be astounded if anyone agrees completely with everything presented here. In the end, any factual errors, mis-interpretations, or misstatements are mine alone.

Notes

Chapter 1. The River, the Valley, and Early Customs with Water

1. Chronic, *Roadside Geology*.
2. The estimates are from Bingham (*The Last Ranch*, 89), and the work of Philip A. Emery, who undertook a number of studies of the San Luis Valley in the 1960s and 1970s for the U.S. Geological Survey and the Colorado Water Conservation Board (Emery, *Hydrology*; Emery et al., *Hydrology*; and Emery et al., *Irrigation*). Abbreviated AF, an acre-foot is the volume of water required to cover an acre (43,560 square feet) to a depth of one foot, or roughly 326,700 gallons. When dealing with enormous volumes of water, such as with irrigation, watershed yields, and reservoirs, acre-feet is more commonly used than gallons.
3. Goetzmann and Williams, *Atlas of North American Exploration*, 146–155.
4. Valdez & Associates, *Culebra River Villages*, E16.
5. Simmons, *San Luis Valley*, 87; Steinel and Working, *History of Agriculture in Colorado*, 177.
6. Pena, "Cultural Landscapes and Biodiversity," 110.
7. Simmons, *San Luis Valley*, 88–89.
8. Rodriguez, *Acequias*, 16.
9. Simmons, *San Luis Valley*, 108.
10. Rodriguez, *Acequias*, 23.
11. Ibid., 115.
12. Crawford, *Mayordomo, Chronicle of an Acequia*, 173.
13. Vandiver interview, March 11, 2014; Colorado Water Conservation Board and Colorado Agricultural and Mechanical College, "Hundred Years of Irrigation in Colorado," 17.
14. Vacquez interview, March 10, 2014.

Chapter 2. Mining and Farming—One Fed the Other

1. Beckwith, "Report of Explorations."
2. Ibid., 40.
3. Ibid., 40.
4. Ibid., 40.
5. Ibid., 41.
6. Ibid., 42.
7. Ibid., 44.
8. Ibid., 45.
9. Ibid., 45.

10. Ubbelohde, Benson, and Smith, *Colorado History*, 61; West, *Contested Plains*, 145.
11. Patton, *Geology and Ore Deposits*, 11.
12. Ibid., 12.
13. Ibid., 66–67.
14. Simmons, *San Luis Valley*, 176–178; Patton, *Geology and Ore Deposits*, 9.
15. Ubbelohde, Benson, and Smith, *Colorado History*, 72–73.

Chapter 3. A Doctrine for Water Takes Shape
1. Stenzel and Cech, *Water: Colorado's Real Gold*, 72.
2. Steinel and Working, *History of Agriculture in Colorado*, 221.
3. Stenzel and Cech, *Water: Colorado's Real Gold*, 184–185.
4. Reich, "'Hispanic' Roots of Prior Appropriation."
5. Stenzel and Cech, *Water: Colorado's Real Gold*, 54.
6. Steinel and Working, *History of Agriculture in Colorado*, 174.
7. Stenzel and Cech, *Water: Colorado's Real Gold*, 72–74; Ubbelohde, Benson, and Smith, *Colorado History*, 189.
8. Hobbs Jr., *Public's Water Resource*, 67.
9. Schorr, *Colorado Doctrine*, 33–36.
10. Steinel and Working, *History of Agriculture in Colorado*, 61; National Park Service, *State by State Numbers*.
11. Ubbelohde, Benson, and Smith, *Colorado History*, 252.
12. Ibid., 61.
13. 1876 Colorado Constitution, Article 16, § 5–6, https://ballotpedia.org/Article _XVI,_Colorado_Constitution.
14. Anderson et al., "Tapping Water Markets."
15. Pisani, "Enterprise and Equity," 16.
16. For those interested in the origin, structure, and modern practice of Colorado water law in greater and excellent detail, the following references are highly recommended and will prove instructive: Jones and Cech, *Colorado Water Law*; Stenzel and Cech, *Water: Colorado's Real Gold*; and Hobbs Jr., *Public's Water Resource*.

Chapter 4. Agriculture Settles In
1. Bartlett, *Great Surveys of the American West*. The four men who led these surveys were Ferdinand V. Hayden, Clarence King, John Wesley Powell, and George Wheeler.
2. Foster, *Strange Genius*, 251.
3. Hayden, *Third Annual Report*, 110.
4. Ibid., 176.
5. Ibid., 246.
6. Thomas, "Agriculture," 195.
7. Ibid., 200.
8. Bartlett, *Great Surveys of the American West*.
9. Wheeler, "Economical Features of Central Colorado," Atlas Sheet No. 61(B); Wheeler, "Land Classification Map of Part of Southwestern Colorado," Atlas Sheet 64(D); Wheeler, "Land Classification Map of Part of Southwestern

Colorado," Atlas Sheet No. 61(D); Wheeler, "Economic Features of Parts of Southern Colorado and Northern New Mexico," Atlas Sheet No. 70(A).

10. Hayden, *Third Annual Report*, 176.
11. Wheeler, "Economic Features of Parts of Southern Colorado," Atlas 70(A).
12. Hayden, *Great West*, 114.
13. Ibid., 115.

Chapter 5. Era of Bonanza Farming

1. Steinel and Working, *History of Agriculture in Colorado*, 70.
2. Ibid., 190–191.
3. Ibid., 187.
4. Colorado Water Conservation Board and Colorado Agricultural and Mechanical College, "Hundred Years of Irrigation in Colorado," 17.
5. Jones and Cech, *Colorado Water Law*, 67.
6. Ogburn, "History of the Development of San Luis," 9.
7. Thomas, "Agriculture," 200.
8. Phillips, Hall, and Black, *Reining in the Rio Grande*, 69.
9. Ogburn, "History of the Development of San Luis," 9.
10. CFS is widely used hydrological shorthand for "cubic feet per second," a rate of flow of about 450 gallons per minute.
11. Ogburn, "History of the Development of San Luis," 10.
12. Phillips, Hall, and Black, *Reining in the Rio Grande*, 68–69.
13. Siebenthal, *Geology and Water Resources*, 26.
14. Stanwyck, *Colorado County History*.
15. Simmons, *San Luis Valley*, 208.
16. Ibid., 159–171.
17. An aquifer is a saturated geological stratum that yields water in sufficient quantity to be economically useful. An unconfined aquifer, sometimes referred to as a water table aquifer, exists where groundwater is exposed to the atmosphere through openings in the overlying geological material. A confined aquifer exists where groundwater is isolated from the atmosphere at the point of discharge by an impermeable geologic formation. The confined aquifer typically is subject to pressures higher than atmospheric pressure; consequently, groundwater wells completed in confined aquifers frequently, though not always, flow at the earth's surface (Driscoll, *Groundwater and Wells*, 62).
18. Paddock, *Introduction to Water Resources*, 18.
19. Phillips, Hall, and Black, *Reining in the Rio Grande*, 69.
20. Paddock, *Introduction to Water Resources*, 1.
21. Phillips, Hall, and Black, *Reining in the Rio Grande*, 84–85.
22. Colorado Department of Local Affairs, *Historical Census Data*.

Chapter 6. The Role and Importance of Storage

1. Smith interview, April 10, 2013.
2. San Luis Valley Irrigation District, "Commemorative Brochure."
3. Montgomery Watson Harza Americas, Inc. Project Team, *Rio Grande Headwaters Restoration*, table 3-9.
4. Follett, *Study of the Use of Water*.

5. Siebenthal, *Geology and Water Resources*, 20.
6. National Resources Committee, *Regional Planning*, 67.
7. Paddock, *Introduction to Water Resources*, 1.
8. Phillips, Hall, and Black, *Reining in the Rio Grande*, 88; U.S. Bureau of Reclamation, "San Luis Valley Project." 3.
9. Paddock, *Introduction to Water Resources*, 2.
10. Vandiver interview, March 11, 2014.
11. San Luis Valley Irrigation District, "Commemorative Brochure."
12. M. Getz and C. Getz, "San Luis Valley"; see also Colorado Foundation for Water Education, "Rio Grande Basin," 6; and Paddock, *Introduction to Water Resources*, 2.
13. Paddock, *Introduction to Water Resources*, 2; see also Colorado Water Conservation Board, *Colorado Statewide Water Supply*.

Chapter 7. Making an Aquifer

1. Smith interview, April 10, 2013.
2. Ibid.
3. Ibid.
4. Worster, *Rivers of Empire*, 313; Opie, *Ogallala*, 142.
5. Siebenthal, *Geology and Water Resources*, 27.
6. Evapotranspiration includes both evaporation and transpiration, the latter being moisture consumed and given off by plants as part of their biological functions. "In most parts of the world, transpiration and ordinary evaporation cannot be differentiated; thus, the loss of water from a land surface is called evapotranspiration" (Driscoll, *Groundwater and Wells*, 54).
7. An aquifer is defined as a geological unit that contains sufficient saturated and permeable material such that it can yield economical quantities of water to wells and springs. Groundwater in an unconfined aquifer is under the pressure exerted by overlying water and atmospheric pressure, whereas groundwater in a confined aquifer, sometimes referred to as an artesian aquifer, is under the pressure of overlying impervious geological layers. See also Driscoll, *Groundwater and Wells*, 61–62.
8. Travis Smith has indicated that the groundwater table three miles east of Center was just eighteen inches deep as recently as 1992.
9. American Meteorological Society, "Glossary." Waterlogged is a "Condition of land where the water table stands at or near the land surface, reaching into the root zone, and may be detrimental to plant growth." Waterlogging is often accompanied by excessive soil salinity, as waterlogged soils prevent leaching of the salts imported by the irrigation water.
10. Siebenthal, *Geology and Water Resources*, 25–26.
11. A common though by no means universal characteristic of wells completed in confined aquifers is their capacity to flow at the ground surface without pumping. Many flowing wells exist in the San Luis Valley.
12. Paddock, *Introduction to Water Resources*, 18.
13. Ibid.
14. American Meteorological Society, "Glossary." Irrigation return flow is "the amount of irrigation water that infiltrates past the root zone and returns fully or partially to the drainage network or system." Where it ultimately ends up

varies with the environment and circumstances. If it collects in a ditch or drain, it can flow by gravity to another irrigator or back to a receiving stream or river. If the latter, it can become available once again for diversion to irrigate crops. The National Resources Committee (*Regional Planning*, 57) concluded that considerable subsurface return flow is re-diverted for irrigation in the San Luis Valley.

Chapter 8. The Compact and the Closed Basin

1. Paddock, *Introduction to Water Resources*, 1–4.
2. Ibid., 5.
3. U.S. Bureau of Land Management, "Rio Grande Compact."
4. Follett, *Study of the Use of Water.*
5. National Resources Committee, *Regional Planning*, 57.
6. U.S. Bureau of Land Management, "Rio Grande Compact."
7. National Resources Committee, *Regional Planning*, 57.
8. Paddock, *Introduction to Water Resources*, 10–11; U.S. Bureau of Land Management, "Rio Grande Compact."
9. Hayden, *Third Annual Report.*
10. Siebenthal, *Geology and Water Resources.*
11. Paddock, *Introduction to Water Resources*, 5.
12. Phreatophytes are plants that obtain their water from shallow groundwater that is within reach of their roots.
13. National Resources Committee, *Regional Planning*, 123.
14. Ibid., 123, 125.
15. Named for Franklin Eddy, the first general manager of the Rio Grande Water Conservation District.
16. U.S. Bureau of Land Management, "Rio Grande Compact."
17. U.S. Bureau of Reclamation, "San Luis Valley Project"; Alamosa Field Division Staff, Bureau of Reclamation, 2013. Ralph Curtis, former general manager of the Rio Grande Water Conservation District, places the Closed Basin Drain's completion in 1992 (interview, April 11, 2013), although Ogburn ("History of the Development of San Luis," 28) indicated that the project was still nearing completion four years later.
18. Bureau of Land Management, "Rio Grande Compact."
19. Paddock, *Introduction to Water Resources*, 23.

Chapter 9. Competing Interests Butt Heads

1. To administer Colorado's water rights, the state legislature long ago divided the state into seven districts. Division 3, with its office in Alamosa, covers the San Luis Valley and upper Rio Grande watershed in Colorado.
2. Montgomery Watson Harza Americas, Inc. Project Team, *Rio Grande Headwaters Restoration*, ES 12–14.
3. Ibid., ES 2 and 2–5.
4. Ibid., fig. 3-2.
5. Ibid., ES-5.
6. Ibid., ES-8.
7. Ibid., ES-15.
8. Ibid., ES-11.

9. Wulf, *Invention of Nature*, 67–68.

10. Ogburn, "History of the Development of San Luis," 30.

11. Paddock, *Introduction to Water Resources*, 25.

12. Bingham, *Last Ranch*, 89.

13. Ibid., 139.

14. Ibid., 171.

15. Ogburn, "History of the Development of San Luis," 30.

16. Bingham, *Last Ranch*.

Chapter 10. A Fire and a Kitchen Sink

1. Smith interview, September 11, 2013.

2. Smith interview, April 10, 2013.

3. Smith interviews September 11, 2013, and April 10, 2013.

4. The Rio Grande Dam rehabilitation project ultimately underwent major changes and upgrades. The Phase One seepage control task was completed in late 2013, roughly on time and within the five-million-dollar budget. But by mid-2015, Phase Two, repairing the dam's outlet structure, had not been started, and its proposed cost had climbed to twenty-six million dollars. Five years later, the rehabilitation project had undergone another massive reformation, including reconstructing the dam's outlet structure, adding a forward guard gate, additional grouting, and improving the spillway's retaining walls. To facilitate this phase, a land exchange was completed between the irrigation district and the adjacent national forest. The district was working with CWCB on a loan-grant package for twenty-five million dollars to complete Phase Two. Ultimately, the district held a special election to authorize a fifteen million dollar loan to complement a ten million dollar CWCB grant. The loan was approved and Phase Two was begun in November 2018 and completed in late 2019. Phase Three was no longer needed, because a new Phase Four will add a bladder to the spillway crest to increase reservoir storage by 10,000 AF. Compared to the initial estimate of twenty-five million dollars for all three phases, reality took a toll on everyone's optimism (Smith interview, May 13, 2016, Smith interview, January 11, 2018; San Luis Valley Irrigation District, Rio Grande Reservoir Rehabilitation).

5. Smith interview, September 11, 2013.

6. Ibid.

7. Ibid.

8. Hobbs Jr., "Colorado Water Law," 14.

9. Colorado Foundation for Water Education, "Rio Grande Basin."

Chapter 11. Crisis in the Valley

1. Vandiver interview, March 15, 2013.

2. Ibid.

3. Ibid.

4. Vandiver interview, March 11, 2014.

5. Leonard Rice Consulting Water Engineers, Inc., "Historic Crop Consumptive Use," fig. 5.

6. A *water year* differs from a calendar year. As defined by the U.S. Geological

Survey, a water year begins October 1 and ends September 30 of the following year. The convention was established in an effort to measure precipitation and runoff during a single meteorological season by including all mountain snowfall from the fall months and attributing it to runoff during the following spring and summer. Water years are named for the year in which they end. Thus, Water Year 2002 began October 1, 2001, and ended September 30, 2002.

7. Leonard Rice Engineers, Inc., "Historic Crop Consumptive Use," figs. 1–3.
8. Ibid., fig. 4.
9. Rio Grande Water Conservation District, "Presentation Concerning Special Improvement"; Davis Engineering Service, Inc., "Change in Unconfined Aquifer."
10. Jones and Cech, *Colorado Water Law*, 120.
11. Montgomery Watson Harza Americas, Inc. Project Team, *Rio Grande Headwaters Restoration*, 2-8.
12. Hobbs Jr., "Colorado Water Law," 7.
13. Hobbs Jr., *Public's Water Resource*, 358.
14. Denison interview, September 2, 2015; Hobbs interview, September 22, 2016.
15. Vandiver interview, May 12, 2016. There appears to be no unanimity among state water officials, however. When specifically queried about this issue, Mike Sullivan, Deputy State Engineer, denied that such an informal water right modification exists (Sullivan interview, June 15, 2017).

Chapter 12. Collaboration as an Economic and Cultural Tool
1. Vandiver interview, March 11, 2014.
2. *Concerning the Matter of the Rules*, Case No. 2004 CW 24, table 6.
3. DiNatale Water Consultants, *Rio Grande Basin Implementation*, 8.
4. Vandiver interview, March 15, 2013.
5. Paddock, *Introduction to Water Resources*, 18.
6. Vandiver interview, March 11, 2014.
7. *Frazier v. Brown* (1861).
8. Vandiver interview, March 15, 2013.
9. *Concerning the Matter of the Rules*, Case No. 2004 CW 24, 86.
10. Vandiver interview, March 15, 2013.
11. Harmon interview, September 16, 2013.
12. Vandiver interview, March 15, 2013.
13. Rio Grande Water Conservation District and Davis Engineering Service, Inc., *Special Improvement District #1*, 2016, table 1.4. Net groundwater consumptive use is estimated as total groundwater pumping modified by credits for surface water recharge and irrigation return flow.
14. Curtis interview, April 11, 2013.
15. Paddock, *Introduction to Water Resources*, 29.
16. Robbins interview, August 15, 2013.
17. Vandiver and Robbins, "Presentation Concerning Special Improvement."
18. *Fallowing* is a practice where farmers temporarily take land out of production for one or more growing seasons. Crops are not planted; irrigation water is not applied.
19. Vandiver and Robbins, "Presentation Concerning Special Improvement."

20. Paddock, *Introduction to Water Resources*, 32; Rio Grande Water Conservation District, "Plan of Water Management," 17.

21. Rio Grande Water Conservation District and Davis Engineering Service, Inc., *Special Improvement District #1*, 2016, fig. 12-1.

22. Vandiver and Robbins, "Presentation Concerning Special Improvement."

23. Simpson interview, January 9, 2018.

24. Ibid.

25. Robbins interview, August 15, 2013.

26. DiNatale Water Consultants, *Rio Grande Basin Implementation*, table 5.

27. Vandiver and Robbins, "Presentation Concerning Special Improvement."

28. Vandiver interview, May 12, 2016.

29. Simpson interview, April 9, 2019.

30. Simpson interview, January 9, 2018.

31. Colorado Revised Statutes 37-92-501(4)(a)(1), https://codes.findlaw.com/co/title-37-water-and-irrigation/co-rev-st-sect-37-92-501.html.

32. Ibid.

33. Jensen et al., *Colorado Acequia Handbook*.

34. Vandiver interview, March 11, 2014.

35. Vacquez interview, March 10, 2014.

Chapter 13. Elsewhere in Colorado and the West

1. U.S. Department of Agriculture, National Agricultural Statistics Service, "Census of Agriculture," 2018, table 1.

2. Limerick and Hanson, *Ditch in Time*, 28–33.

3. Ibid., 32.

4. Steinel and Working, *History of Agriculture in Colorado*, 53.

5. Ibid., 202.

6. CDM, *SWSI 2010 South Platte Basin Report*, 8-8 to 8-14.

7. Steinel and Working, *History of Agriculture in Colorado*, 226.

8. CDM, *SWSI 2010 South Platte Basin Report*, table 4-8.

9. Coleman, *Citizen's Guide to Colorado's Transbasin*, 5.

10. Steinel and Working, *History of Agriculture in Colorado*, 223–226.

11. Hydrographic Branch, "Map of Colorado Historic Average."

12. CDM, *SWSI 2010 South Platte Basin Report*, 8-8 to 8-12.

13. Waskom, "Tale of Two Rivers."

14. CDM, *SWSI 2010 South Platte Basin Report*, 8-8.

15. Colman, "America's Biggest Water Users."

16. CDM, *SWSI 2010 South Platte Basin Report*, fig. 4-6.

17. Steinel and Working, *History of Agriculture in Colorado*, 16.

18. MacDonnell, *From Reclamation to Sustainability*, 23.

19. Ibid., 26.

20. Colorado Water Conservation Board and Colorado Agricultural and Mechanical College, "Hundred Years of Irrigation in Colorado," 15–16.

21. MacDonnell, *From Reclamation to Sustainability*, 26–30.

22. Steinel and Working, *History of Agriculture in Colorado*, 215–219.

23. MacDonnell, *From Reclamation to Sustainability*, 40–41.

24. Ibid., 40–44.

25. Colorado Division of Water Resources, "Arkansas Basin Fact Sheet."

26. CDM, *SWSI 2010 South Platte Basin Report*, table 4-10.
27. Hydrographic Branch, "Map of Colorado Historic Average."
28. Coleman, *Citizen's Guide to Colorado's Transbasin*, 19.
29. Ibid., 70.
30. MacDonnell, *From Reclamation to Sustainability*, 76.
31. Ibid., 240.
32. Coleman, *Citizen's Guide to Colorado's Transbasin*, 9.
33. CDM, *SWSI 2010 South Platte Basin Report*, fig. 4-6.
34. Colorado Division of Water Resources, "Arkansas Basin Fact Sheet."
35. In addition to Colorado, the Intermountain West as herein defined includes Wyoming, Idaho, western Montana, Utah, New Mexico, Arizona, Nevada, and the eastern halves of Oregon and Washington. A sliver of eastern California is included in some definitions of this region, but California is not examined here. California's agricultural economy is huge and its water laws are complex. California adopted both riparian and prior-appropriation water rights, and its courts subsequently added unique variations that differ from other western states. For those readers wishing a contemporary treatment of California agriculture and the environment, I recommend *The Dreamt Land* by Mark Arax.
36. Colorado State Constitution, Article XVI, Section 6, https://ballotpedia.org/Article_XVI,_Colorado_Constitution.
37. Hobbs Jr., *Public's Water Resource*, 359.
38. Maupin et al., *Estimated Use of Water*, table 7.
39. Ibid., 25.
40. Holub, "Summary of Arizona Water," 1; James and O'Dell, "Megafarms and Deeper Wells."
41. James and O'Dell, "Megafarms and Deeper Wells."
42. The number of irrigated acres for each major irrigation method was multiplied by the application rate in AF/ACRE and an accepted irrigation efficiency percentage for both methods. According to Mike Sullivan, Colorado's Deputy State Engineer (2017), sprinkler irrigation is estimated to have an average efficiency of 83 percent and flood irrigation 50 percent.
43. Maupin et al., *Estimated Use of Water*, table 7.
44. Richter et al., "Water Scarcity and Fish."
45. MacDonnell, *From Reclamation to Sustainability*, 241.

Chapter 14. A New Normal

1. Barbier, *The Water Paradox*, 1.
2. Ibid., 23.
3. Ibid., 24.
4. Jones and Cech, *Colorado Water Law*, 103.
5. Rio de la Vista interview, April 2, 2019.
6. Powell and Mutz, *Ground-Water Resources of the San Luis Valley*, 57.
7. Paddock, *Introduction to Water Resources*, 11.
8. The Environmental Quality Incentives Program (EQIP), administered by the Natural Resources Conservation Service, a branch of the U.S. Department of Agriculture, is the federal agency most involved in the financial support of irrigators pursuing modernization.
9. Simpson interview, January 9, 2018.

10. M. Sullivan interview, June 15, 2017.

11. Hobbs Jr., "Colorado Water Law," 14.

12. San Luis Valley Irrigation District, "Rio Grande Reservoir Rehabilitation."

13. Bunge and Newman, "To Stay on the Land."

14. M. Sullivan interview, June 15, 2017.

15. Barbier, *The Water Paradox*, 151.

Epilogue

1. Simpson interview, January 9, 2018.

2. Three months later, in his April 2018 notice to the San Luis Valley irrigation community, Cotten reduced his predicted 2018 Rio Grande runoff to 300,000 AF. For calendar year 2018, the Rio Grande eventually delivered about 280,000 AF to the valley, or less than half of the river's average yield.

3. Dettinger, Udall, and Georgakakos, "Western Water and Climate."

4. Simpson interview, April 9, 2019.

5. Rio Grande Water Conservation District, "Plan of Water Management."

6. Simpson, "Tale of Two Rivers."

7. Purtell, "As Metro Denver Grows."

8. Ibid.; J. Smith, "Denver Developer."

9. Cotten, "How the Rio Grande Works."

10. Simpson, "Tale of Two Rivers."

11. Cotten, *Rio Grande Compact*.

12. Bowlin, "Colorado Farmers Fight."

13. Davis Engineering Service, Inc., "Change in Unconfined Aquifer."

References

American Meteorological Society. "Glossary of Meteorological Terms." 2015.

Anderson, Terry L., Donald R. Leal, Brandon Scarborough, and Lawrence Reed Watson. "Tapping Water Markets." In *Free Market Environmentalism for the Next Generation*, by Terry L. Anderson and Donald R. Leal. New York: Palgrave Macmillan, 2015.

Anderson, Terry L., Donald R. Leal, and Shawn Regan. "Who Owns the Environment?" In *Free Market Environmentalism for the Next Generation*, by Terry L. Anderson and Donald R. Leal. New York: Palgrave Macmillan, 2015.

Arax, Mark. *The Dreamt Land*. New York: Knopf, 2019.

Barbier, Edward B. *The Water Paradox: Overcoming the Global Crisis in Water Management*. New Haven, CT: Yale University Press, 2019.

Bartlett, Richard A. *Great Surveys of the American West*. Norman, OK: University of Oklahoma Press, 1962.

Beckwith, E. G., Lt. "Report of Explorations for a Route for the Pacific Railroad, by Capt. J. W. Gunnison, Topographical Engineers Near the 38th and 39th Parallels of North Latitude from the Mouth of the Kansas River, Mo., to the Sevier Lake, in the Great Basin," 1855. *Making of America Books*. Accessed Jan. 17, 2013. http://quod.lib.umich.edu/m/moa/afk4383.0002.001/1?view=pdf.

Benson, Reed D. "We Should Recognize the Legal Rights of Rivers." *High Country News*, Dec. 13, 2017.

Bingham, Sam. *The Last Ranch: A Colorado Community and the Coming Desert*. San Diego, CA: Harcourt, Brace & Company, 1996.

Borunda, Alejandra. "How Beef Eaters in Cities Are Draining Rivers in the American West." *National Geographic*, March 2, 2020. Accessed March 2, 2020. https://www.nationalgeographic.com/science/2020/03/burger-water-shortages-colorado-river-western-us/?utm_campaign=Rockies%20Today&utm_medium=email&utm_source=Revue%20newsletter#close.

Bowlin, Nick. "Colorado Farmers Fight to Save Their Water and Their Community's Future." *High Country News*, September 16, 2019.

———. "A Water 'Win-Win' in Colorado? Not So Fast." *High Country News*, September 16, 2019.

Brink, Phil, and Greg Peterson. "Water and Agriculture: Valuing Our Essential Resources." *Colorado Cattlemen's Association Ag Water Network*, April 2020. Accessed 2020. www.agwaternetwork.org.

Brocious, Ariana. "One Irrigator's Waste is Another's Supply." *Western Confluence*, Dec. 23, 2014. http://www.westernconfluence.org/one-irrigators-waste-is-anothers-supply/.

Bunge, Jacob, and Jesse Newman. "To Stay on the Land, American Farmers Add Extra Jobs." *The Wall Street Journal*, Feb. 25, 2018.

Business Water Task Force. "New Mexico Water Basics and an Introduction to Water Markets." *Campanastan*, 2010. Accessed Feb. 25, 2020. https://aquadoc.typepad.com/files/nm-water-brochure-final.pdf.

CDM. *SWSI 2010 South Platte Basin Report, Basinwide Consumptive and Nonconsumptive Water Supply Needs Assessment*, final draft report, 2011. Denver: Colorado Water Conservation Board.

Chronic, Halka. *Roadside Geology of Colorado*. Missoula, MT: Mountain Press Publishing Co., 1980.

Cohen, Paul E. *Mapping the West: America's Westward Movement 1524–1890*. New York: Rizzoli International Publications, 2002.

Coleman, Caitlin. *Citizen's Guide to Colorado's Transbasin Diversions*. Denver: Colorado Foundation for Water Education, 2014.

Colman, Zack. "America's Biggest Water Users—Farmers—Learn to Use Less of It." *The Christian Science Monitor*, March 3, 2017.

———. "How Water Swaps Help the West Manage a Precious Resource." *Christian Science Monitor*, March 8, 2017.

Colorado Agricultural Water Alliance. "Can Agricultural Water Conservation and Efficiency Provide the Water Needed for Colorado's Future?" *Colorado Ag Water Alliance*, May 13, 2018. https://docs.wixstatic.com/ugd/302b62_a42489b a652d4385b706692a46117769.pdf.

Colorado Department of Local Affairs. *Historical Census Data—Counties and Municipalities*, 2017. Accessed April 3, 2017. demography.dola.coloradogov/population/data/historical_census.

Colorado Division of Water Resources. "Arkansas Basin Fact Sheet." *Statewide Water Supply Initiative*. Denver: Colorado Department of Natural Resources, Feb. 2006.

Colorado Foundation for Water Education. "Rio Grande Basin." *Headwaters* (fall 2005): 6–7.

Colorado Water Conservation Board. *Colorado Statewide Water Supply Initiative Fact Sheet: Major Storage Projects in the Rio Grande Basin*. Denver: Colorado Water Conservation Board, 2006.

———. *Colorado's Water Plan*. Denver, CO: Colorado Water Conservation Board, 2015.

Colorado Water Conservation Board and Colorado Agricultural and Mechanical College. "A Hundred Years of Irrigation in Colorado: 100 Years of Organized and Continuous Irrigation 1852–1952," *Colorado Water Conservation Board, Department of Natural Resources*, 1952. Accessed Jan. 4, 2019. https://dnrweb link.state.co.us/cwcbsearch/DocView.aspx?id=26331&searchid=cfaf93fa-7387 -4d83-836d-51929393d06e&dbid=0.

Concerning the Matter of the Rules Governing New Withdrawals of Ground Water in Water Division No. 3 Affecting the Rate or Direction of Movement of Water in the Confined Aquifer System, aka "Confined Aquifer New Use Rules for Division 3." Case No. 2004 CW 24. District Court, Water Division No. 3, Colorado, November 9, 2006. https://www.courts.state.co.us/Courts/Water/Rulings /Div3/04CW24%20Part%20I-IV.pdf.

Cotten, Craig, Division 3 Engineer. "How the Rio Grande Works." *State of the Basin Symposium, Adams State University,* 2019. Alamosa, CO: Salazar Rio Grande del Norte Center & Department of Biology and Earth Sciences. Accessed March 2019. https://www.youtube.com/playlist?list=PLM1XIDdQr4T5uncIUerKvQU hESIzAcfoO.

———. *Rio Grande Compact, March 11, 2019 Analysis,* 2019. 10-day report, Alamosa, CO: Division 3, Colorado Division of Water Resources.

Crawford, Stanley. *Mayordomo: Chronicle of an Acequia in Northern New Mexico.* Albuquerque, NM: University of New Mexico Press, 1988.

Curtis, Ralph, General Manager (Retired), Rio Grande Water Conservation District. Interview by D. Stiller, April 11, 2013.

Davis Engineering Service, Inc. "Change in Unconfined Aquifer Storage, Year 2002–2020." *Rio Grande Water Conservation District,* July 9, 2020. Accessed July 27, 2020. https://rgwcd.org/well-information.

de la Vista, Rio, Director, Salazar Rio Grande del Norte Center, Adams State University. Interview by D. Stiller, April 2, 2019.

Denison, Cary, Former Water Commissioner, Division 4, Colorado State Engineer's Office. Interview by D. Stiller, September 2, 2015.

Dettinger, Michael, Bradley Udall, and Aris Georgakakos. "Western Water and Climate Change." *Ecological Applications* 25(8) (2015): 2069–2093.

DeVoto, Bernard. *The Year of Decision 1846.* New York: Little, Brown, 1942.

DiNatale Water Consultants. *Rio Grande Basin Implementation Plan (Draft).* Alamosa, CO: Rio Grande Basin Roundtable, 2014.

Disappearing West. "The Disappearing West: Rivers." *The Disappearing West,* Feb. 12, 2018. https://disappearingwest.org/rivers.html.

Driscoll, Fletcher G. *Groundwater and Wells.* 2nd ed. St. Paul, MN: Johnson Division, 1986.

Emery, Philip A. *Hydrology of the San Luis Valley in South-Central Colorado.* Hydrologic Atlas 381, Washington, D.C.: U.S. Geological Survey, 1971.

Emery, Philip A., A. J. Boettcher, R. J. Snipes, and H. J. McIntyre Jr. *Hydrology of the San Luis Valley, South-Central Colorado.* Open-File Report, Washington, D.C.: U.S. Geological Survey, in cooperation with the Colorado Water Conservation Board, 1969.

Emery, Philip A., John M. Dumeyer, and Harold J. McIntyre. *Irrigation and Municipal Wells in the San Luis Valley, Colorado.* Open-File Report 70–119, Washington, D.C.: U.S. Geological Survey, 1969.

Environmental Law Foundation. "California's Public Trust Doctrine." *envirolaw.org,* Jan. 18, 2018. https://www.envirolaw.org/documents/ScottFAQ.pdf.

Fiege, Mark. *Irrigated Eden: The Making of an Agricultural Landscape in the American West.* Seattle: University of Washington Press, 1999.

Follett, W. W. *A Study of the Use of Water for Irrigation on the Rio Grande del Norte Above Fort Quitman, Texas.* El Paso, TX: U.S. Army Corps of Engineers, 1896.

Foster, Mike. *Strange Genius: The Life of Ferdinand Vandeveer Hayden.* Niwot, CO: Roberts Rinehart Publishers, 1994.

Fowler, Jacob. *The Journal of Jacob Fowler, Narrating an Adventure from Arkansas Through the Indian Territory, Oklahoma, Kansas, Colorado, and New Mexico, to the Sources of Rio Grande del Norte, 1821–1822.* Ed. Elliott Coues. New York: Francis P. Harper, 1898.

Fradkin, Philip L. *A River No More: The Colorado River and the West.* Expanded and updated ed. Berkeley, CA: University of California Press, 1996.

Frazier v. Brown. 12 Ohio St. Ohio Supreme Court, 1861.

Gardner-Smith, Brent. "State of Colorado's Instream Flow Program Is Lauded, Challenged." *Aspen Daily News,* Jan. 21, 2014.

Getz, Melvin, and Camille Getz. "San Luis Valley Will Celebrate a Century of Reservoirs." *Colorado Central Magazine,* July 1, 2007. Accessed August 6, 2014. http://cozine.com/2007-july/San-luis-valley-will-celebrate-a-century-of-reservoirs.

Goetzmann, William H., and Glyndwr Williams. *The Atlas of North American Exploration: From the Norse Voyages to the Race to the Pole.* Norman, OK: University of Oklahoma Press, 1998.

Harmon, Eric J., Principal, HRS Water Consultants, Inc. Interview by D. Stiller, September 16, 2013.

Harvey, Nelson. "A Price for the Priceless: How Do We Value Colorado's Water?" *Headwaters,* summer 2016: 20–26.

Hayden, Ferdinand V. *The Great West: Its Attractions and Resources.* Bloomington, IL: Charles R. Brodix, 1880.

———. *Preliminary Report of the United States Geological Survey of Wyoming and Portions of Contiguous Territories.* Washington, D.C.: Government Printing Office, 1871.

———. *Third Annual Report of the United States Geological Survey of the Territories, Embracing Colorado and New Mexico.* Washington, D.C.: Government Printing Office, 1869.

Heide, Ruth. "Valley Water Export Plan Presented." *Alamosa News,* Dec. 7, 2018. Accessed Dec. 8, 2018. https://alamosanews.com/article/valley-water-export-plan-presented.

Hill, William E. *The Santa Fe Trail: Yesterday and Today.* Caldwell, ID: The Caxton Printers, Ltd., 1992.

Hobbs, Gregory J., Jr. "Colorado Water Law: An Historical Overview." *Water Law Review* vol. 1 (fall 1997).

———. *The Public's Water Resource: Articles on Water Law, History, and Culture.* Denver, CO: Continuing Legal Education in Colorado, Inc., 2007.

Hobbs, Gregory J., Sturm College of Law, University of Denver. Interview by D. Stiller, September 22, 2016.

Holub, Hugh. "Summary of Arizona Water Law." *Groundwater Awareness League,* Oct. 2009. Accessed Jan. 13, 2020. http://www.g-a-1.info/Water-Law.htm.

Horgan, Paul. *Great River: The Rio Grande in North American History.* Middletown, CT: Wesleyan University Press, 1984.

Hydrographic Branch, State Engineer's Office, Colorado Division of Water Resources. "Map of Colorado Historic Average Annual Stream Flows." Denver, CO: Division of Water Resources, Colorado Department of Natural Resources, 2011.

Idaho Department of Water Resources. "A Water Users Information Guide: Idaho Water Rights, a Primer." *Water Rights,* July 2015. Accessed Janary 2020. https://idwr.idaho.gov/files/water-rights/water-rights-brochure.pdf.

James, Edwin. *Account of an Expedition from Pittsburgh to the Rocky Mountains, Performed in the Years 1819 and 1820*. Philadelphia: H.C. Carey and I. Lea, 1823.

James, Ian, and Rob O'Dell. "Megafarms and Deeper Wells Are Draining the Water Beneath Rural Arizona—Quietly, Irreversibly." *azcentral*, Dec. 27, 2019. Accessed Jan. 11, 2020. https://www.azcentral.com/in-depth/news/local/arizona-environment/2019/12/05/unregulated-pumping-arizona-groundwater-dry-wells/2425078001/.

Jensen, Jens, Peter Nichols, Ryan Golten, Sarah Krakoff, Sarah Parmar, Karl Kumli, and Jesse Heibel. *Colorado Acequia Handbook: Water Rights and Governance Guide for Colorado's Acequias* (revised). Boulder, CO: Sangre de Cristo Acequia Association, Getches-Wilkinson Center for Natural Resources, Energy, and the Environment, Colorado Open Lands, and private attorneys, 2016.

Jones, Andrew P., and Tom Cech. *Colorado Water Law for Non-Lawyers*. Boulder, CO: University Press of Colorado, 2009.

Kenney, Doug. "The Colorado River Is Not a Water Buffet. So Why the 'First Come, First Serve' Policy?" *The Guardian*, June 22, 2015. https://www.theguardian.com/commentisfree/2015/jun/16/colorado-river-water-buffet-first-come-first-serve-drought.

Laflin, Rose. *Irrigation, Settlement, and Change on the Cache La Poudre River*. Special Report No. 15, Fort Collins, CO: Colorado Water Resources Research Institute, Colorado State University, 2005.

Leonard Rice Consulting Water Engineers, Inc. "Historic Crop Consumptive Use Analysis—Rio Grande Decision Support System." *Colorado Department of Natural Resources*, June 2004. Accessed March 2019. https://dnrweblink.state.co.us/cwcbsearch/DocView.aspx?id=123306&searchid=45652f5c-4241-4e33-aeac-4430ceaaa2fc&dbid=0.

Limerick, Patricia Nelson. *Legacy of Conquest*. New York: W. W. Norton & Company, Inc., 1987.

Limerick, Patricia Nelson, and Jason L. Hanson. *A Ditch in Time: The City, the West, and Water*. Golden, CO: Fulcrum Publishing, 2012.

Lyell, Charles. *Principles of Geology*. London: John Murray, 1832.

MacDonnell, Lawrence J. *From Reclamation to Sustainability: Water, Agriculture, and the Environment in the American West*. Niwot, CO: University Press of Colorado, 1999.

Maupin, Molly A., Joan F. Kenny, Susan S. Hutson, John K. Lovelace, Nancy L. Barber, and Kristin S. Linsey. *Estimated Use of Water in the United States in 2010*. Circular 1405, Reston, VA: U.S. Geological Survey, 2014.

Montgomery Watson Harza Americas, Inc. Project Team. *Rio Grande Headwaters Restoration Project*. Final report, Alamosa, CO: San Luis Valley Water Conservancy District's Rio Grande Restoration Project Enterprise Technical Advisory Committee, 2001.

Most, Stephen. *River of Renewal: Myth and History in the Klamath Basin*. Portland, OR: Oregon Historical Society Press, 2006.

National Park Service. *State by State Numbers—Homestead National Monument of America*, 2013. Accessed Feb. 21, 2013. http://www.nps.gov/home/historyculture/statenumbers.htm.

National Resources Committee. *Regional Planning: Part VI—The Rio Grande Joint Investigation in the Upper Rio Grande Basin in Colorado, New Mexico, and Texas, 1936–1937.* Washington, D.C.: U.S. Government Printing Office, 1938.

Nicla, Andrew. "Does Arizona Really Use Less Water Now Than It Did in 1957?" *azcentral,* Feb. 12, 2019. Accessed Jan. 11, 2020. https://www.azcentral.com /story/news/local/arizona-environment/2019/02/12/arizona-water-usage-state -uses-less-now-than-1957/2806899002/.

Nugent, Walter. *Habits of Empire: A History of American Expansion.* New York: Vintage Books, 2009.

O'Dell, Rob, and Ian James. "Arizona Has Tried to Safeguard Groundwater Beneath Its Big Cities. But Things Are About to Change." *azcentral,* Dec. 11, 2019. Accessed 2020. https://www.azcentral.com/in-depth/news/local/arizona -environment/2019/12/05/arizona-groundwater-rules-water-tables-declining -parts-phoenix.

Office of the State Engineer. "State Engineer's Statement of Basis and Purpose for Rules Governing the Withdrawal of Groundwater in Water Division No. 3 (The Rio Grande Basin) and Establishing Criteria for the Beginning and End of the Irrigation Season in Water Division No. 3." *Colorado Department of Natural Resources, Water Resources Division, Division 3: Rio Grande River Basin Rules (SLV),* n.d. Accessed Dec. 12, 2019. http://water.state.co.us/DWRIPub/Docu ments/Final%20Statement%20of%20Basis%20and%20Purpose.pdf.

Ogburn, Robert W. "A History of the Development of San Luis Valley Water." *The San Luis Valley Historian* (The San Luis Valley Historical Society) 28, 1 (1996): 5–40.

Olalde, Mark. "After the Gold Is Gone—Rockies Today for Monday, March 18." *Mountain West News,* March 18, 2019. Accessed March 18, 2019. https://mount ainwestnews.org/after-the-gold-is-gone-c81e2a7f7970.

Opie, John. *Ogallala: Water for a Dry Land.* 2nd edition. Lincoln, NE: University of Nebraska Press, 2000.

Oregon Water Resources Department. *Water Rights in Oregon: An Introduction to Oregon's Water Laws.* Salem, OR: State of Oregon, 2018.

Owen, David. "Where the River Runs Dry: The Colorado and America's Water Crisis." *The New Yorker,* May 25, 2015: 52–63.

Paddock, William A. *Introduction to Water Resources Issues in Water Division No. 3, The Rio Grande Basin.* Denver, CO: Carlson, Hammond & Paddock, LLC, 2014.

Paskus, Laura. "As New Mexico Reservoirs Hit Bottom, Worries Grow Over the Future." *News Deeply,* September 25, 2018. Accessed September 28, 2018. https://www.newsdeeply.com/water/articles/2018/09/25/as-new-mexico-reser voirs-hit-bottom-worries-grow-over-the-future.

Patton, Horace B. *Geology and Ore Deposits of the Platoro-Summitville Mining District, Colorado.* Bulletin 13, Boulder, CO: Colorado Geological Survey, 1917. Accessed Jan. 15, 2013.

Pena, Devon G. "Cultural Landscapes and Biodiversity: The Ethnoecology of an Upper Rio Grande Watershed Commons." In *Ethnoecology: Studied Knowledge/ Located Lives.* Tucson, AZ: University of Arizona Press, 1999, 107–132.

Phillips, Fred M., G. Emlen Hall, and Mary E. Black. *Reining in the Rio Grande: People, Land, and Water.* Albuquerque, NM: University of New Mexico Press, 2011.

Pike, Zebulon Montgomery. *The Expeditions of Zebulon Montgomery Pike to Head-*

waters of the Mississippi River, through Louisiana Territory, and in New Spain, During the Years 1805–6–7. Reprint of original 1895 edition. Ed. Elliott Coues. Vol. 2. Minneapolis, MN: Ross & Haines, 1965.

Pisani, Donald J. "Enterprise and Equity: Critique of Western Water Law in the Nineteenth Century." *The Western Historical Quarterly* (Jan. 1987): 15–37.

Powell, William James, and Philip B. Mutz. *Ground-Water Resources of the San Luis Valley, Colorado, with a Section on an Inflow-Outflow Study of the Area.* USGS Water Supply Paper 1379, Washington, D.C.: U.S. Geological Survey, 1958.

Purtell, Joe. "As Metro Denver Grows, Another Caller Wants to Tap the Vast Aquifer Under the San Luis Valley." *The Colorado Sun Annual Report,* October 7, 2019. Accessed Oct. 7, 2019. https://coloradosun.com/2019/10/07/sean-tonner-san-luis-valley-water-pipeline/.

Reich, Peter L. "The 'Hispanic' Roots of Prior Appropriation in Arizona." *Arizona State Law Journal* 1995: 649–662.

Reisner, Marc. *Cadillac Desert: The American West and Its Disappearing Water.* New York: Viking Penguin, 1986.

Rennicke, Jeff. *The Rivers of Colorado.* Vol. 1 of Colorado Geographic Series. Helena, MT: Falcon Press Publishing Co, 1985.

Richter, Brian D., Dominique Bartak, Peter Caldwell, Kyle Frankel Davis, Peter Debaere, Arjen Y. Hoekstra, Tianshu Li, et al. "Water Scarcity and Fish Imperilment Driven by Beef Production." *Nature Sustainability,* March 2020(3): 319–328. Accessed 2020. https://doi.org/10.1038/s41893-020-0483-z.

Rio Grande Water Conservation District. "Plan of Water Management, Amended, Special Improvement District No. 1 of the Rio Grande Water Conservation District," June 6, 2017. Accessed August 1, 2020. https://rgwcd.org/attachments /subdistrict1/Plan%20Revisions/Plan%20Water%20Management%20AMEN DED%206Jun17_draft.pdf.

———. "Presentation Concerning Special Improvement District No. 1 of the Rio Grande Water Conservation District for Consideration by the Water Resources Review Committee of the Colorado Legislature," Sept. 26, 2013. Alamosa, CO: Rio Grande Water Conservation District.

Rio Grande Water Conservation District and Davis Engineering Services, Inc. *Special Improvement District #1 of the Rio Grande Water Conservation District, Annual Report for the 2013 Plan Year.* Alamosa, CO: Rio Grande Water Conservation District, 2014.

———. *Special Improvement District #1 of the Rio Grande Water Conservation District, Annual Report for the 2015 Plan Year.* Alamosa, CO: Rio Grande Water Conservation District, 2016.

Rivera, Jose A. *Acequia Culture: Water, Land, and Community in the Southwest.* Albuquerque, NM: University of New Mexico Press, 1998.

Robbins, David, Attorney for the Rio Grande Water Conservation District. Interview by D. Stiller, August 15, 2013.

Rodriguez, Sylvia. *Acequias: Water-Sharing, Sanctity, and Place.* Santa Fe, NM: School of American Research Press, 2007.

Roscoe, Gerald. *Westward: The Epic Crossing of the American Landscape.* New York: Monacelli Press, 1995.

Sangre de Cristo Acequia Association. *Sangre de Cristo Acequia Association,* 2019. Accessed Dec. 13, 2019. https://www.coloradoacequias.org/.

San Luis Valley Irrigation District. "Commemorative Brochure, Rio Grande Reservoir 100th Anniversary Celebration." Center, CO: San Luis Valley Irrigation District, Aug. 23, 2012.

———. "Rio Grande Reservoir Rehabilitation Project." Center, CO: San Luis Valley Irrigation District, November 5, 2019. Accessed July 28, 2020. https://www.slvid .org/project.

Schorr, David. *The Colorado Doctrine: Water Rights, Corporations, and Distributive Justice on the American Frontier*. New Haven, CT: Yale University Press, 2012.

Scott, Glenn R. *Historic Trail Maps of the Pueblo 1° X 2° Quadrangle, Colorado*. Miscellaneous Investigations Series Map I-930: U.S. Geologial Survey, 1975.

Sibley, George. *Water Wranglers: The 75-Year History of the Colorado River District: A Story About the Embattled Colorado River and the Growth of the West*. Grand Junction, CO: Colorado River Water Conservation District, 2012.

Siebenthal, C. E. *Geology and Water Resources of the San Luis Valley, Colorado*. U.S. Geological Survey Water-Supply Paper 240, Washington, D.C.: U.S. Government Printing Office, 1910.

Simmons, Virginia McConnell. *The San Luis Valley: Land of the Six-Armed Cross*. Second edition. Niwot, CO: University Press of Colorado, 1999.

Simpson, Cleave. "Communities Handling Scarcity." *Ninth Annual Upper Colorado River Basin Water Forum: Tools for Adaptation*. Grand Junction, CO: Colorado Mesa University and Ruth Powell Hutchins Water Center, 2019.

———. "A Tale of Two Rivers: Groundwater Management in the South Platte and Rio Grande Basins." *State of the Basin Symposium, Adams State University*. Alamosa, CO: Salazar Rio Grande del Norte Center & Department of Biology and Earth Sciences, 2019. Accessed March 2019. https://www.youtube.com/play list?list=PLM1XIDdQr4T5uncIUerKvQUhESIzAcfoO.

Simpson, Cleave, General Manager, Rio Grande Water Conservation District. Interview by D. Stiller, January 9, 2018.

———. Interview by D. Stiller, April 9, 2019.

Smith, Jerd. "Denver Developer, Former Governor Make $118M Play for San Luis Valley Water." *Water Education Colorado*, June 26, 2019. Accessed August 6, 2020. https://www.watereducationcolorado.org/fresh-water-news/denver-dev eloper-former-governor-make-118m-play-for-san-luis-valley-water/.

———. "Drought Slams $15M Voluntary Conservation Effort in Colorado Potato Country." *Water Education Colorado*, September 26, 2019. Accessed August 6, 2020. https://www.watereducationcolorado.org/fresh-water-news/drought -slams-15m-voluntary-conservation-effort-in-colorado-potato-country/.

———. "Report: Colorado's Farm Water Use Exceeds National Average, Despite Efforts to Conserve." *Fresh Water News*, Feb. 19, 2020. Accessed Feb. 19, 2020. https://www.watereducationcolorado.org/fresh-water-news/report-colorados -farm-water-use-exceeds-national-average-despite-efforts-to-conserve/.

Smith, Travis, Senior Consultant, DiNatale Water Consultants. Interview by D. Stiller, January 11, 2018.

Smith, Travis, Superintendent, San Luis Valley Irrigation District. Interview by D. Stiller, April 10, 2013.

———. Interview by D. Stiller, September 11, 2013.

———. Interview by D. Stiller, May 13, 2016.

Stanwyck, Don. *Colorado County History*, 2009. Accessed August 14, 2014. http://rootsweb.ancestry.com/~cotttp/coplaces/county-div.html.

Steinel, Alvin T., and D. W. Working. *History of Agriculture in Colorado, 1858 to 1926*. Fort Collins, CO: State Agricultural College, 1926.

Stenzel, Richard L., and Tom Cech. *Water, Colorado's Real Gold: A History of the Development of Colorado's Water, the Prior Appropriation Doctrine and the Colorado Division of Water Resources*. Denver: Richard Stenzel, 2013.

Stiller, David. *Wounding the West: Montana, Mining, and the Environment*. Lincoln, NE: University of Nebraska Press, 2000.

Sullivan, Mike, Deputy State Engineer. Interview by D. Stiller, June 15, 2017.

Sullivan, Vernon L. *Irrigation in New Mexico*. U.S. Department of Agriculture, Office of Experiment Stations, Bulletin 215, Washington, D.C.: Government Printing Office, 1909.

"Summary of Compacts and Litigation Governing Colorado's Use of Interstate Streams." *Colorado Water Conservation Board—Arkansas Basin Roundtable*. CDM, 2000. Accessed Jan. 11, 2019. cwcb.state.co.us/public-information/publications/ReportsStudies/SWSIAppendices/Appendix%20D%20Summary%20of%20Compact%20and%20Litigation.pdf.

Terry, Kevin. Email to David Stiller regarding Colorado Trout Unlimited and Rio Grande Water. Alamosa, Colorado, Jan. 16, 2018.

Thomas, Cyrus. "Agriculture." In *Preliminary Report of the United States Geological Survey of Wyoming and Portions of Contiguous Territories*, by Ferdinand V. Hayden, 188–263. Washington, D.C.: Government Printing Office, 1871.

Trout Unlimited. *Landowner's Guide to Washington Water Rights*. Third edition. Wenatchee, WA: Trout Unlimited—Washington Water Project, 2019.

U.S. Bureau of Land Management. "The Rio Grande Compact and the Role of the Closed Basin Project," n.d. Accessed Jan. 8, 2014. In author's file.

U.S. Bureau of Reclamation. "San Luis Valley Project." *Project Details—San Luis Valley Project—Bureau of Reclamation*, May 17, 2011. Accessed Oct. 21, 2014. http://www.usbr.gov/projects/Project.jsp?proj_Name=San+Luis+Valley+Project.

U.S. Department of Agriculture. *Colorado Agricultural Statistics 2018*. Annual Bulletin. Lakewood, CO: U.S. Department of Agriculture, National Agricultural Statistics Service, and Colorado Department of Agriculture, 2018.

———. Economic Research Service. "State Fact Sheet: Colorado." *State Data*, Nov. 29, 2019. Accessed Jan. 13, 2020. https://data.ers.usda.gov/reports.aspx?StateFIPS=08&StateName=Colorado&ID=17854#.

———. Census of Agriculture. *2018 Irrigation and Water Management Survey*. Nov. 2018. Accessed Jan. 13, 2020. https://www.nass.usda.gov/Publications/AgCensus/2017/Online_Resources/Farm_and_Ranch_Irrigation_Survey/index.php.

———. National Agricultural Statistics Service. Census of Agriculture. "2007 Irrigation and Water Management Survey." *2007 Census of Agriculture*. 2008. Accessed Jan. 13, 2020. https://www.nass.usda.gov/Publications/AgCensus/2007/Full_Report/Volume_1,_Chapter_1_US/usv1.pdf.

Ubbelohde, Carl, Maxine Benson, and Duane A. Smith. *A Colorado History*. 7th ed. Boulder, CO: Pruett Publishing Company, 1995.

Udall, Brad, and Jonathan Overpeck. "Climate Change Is Shrinking the Colorado River." *The Conversation*, June 13, 2017. http://theconversation.com/climate-change-is-shrinking-the-colorado-river-76280.

Utah Office of Legislative Research and General Counsel. "How Utah Water Works." *Utah Division of Water Rights*, Nov. 2012. Accessed Feb. 26, 2020. https://www.waterrights.utah.gov/wrinfo/Brochures/how_utah_water_works.pdf.

Vacquez, Eugene, Costilla County farmer. Interview by D. Stiller, March 10, 2014.

Valdez & Associates. *The Culebra River Villages of Costilla County, Colorado*. National Register of Historic Places, Washington, DC: U.S. Department of the Interior, National Park Service, 2000.

Vandiver, Steve, and David Robbins. "Presentation Concerning Special Improvement District No. 1 of the Rio Grande Water Conservation District for Consideration by the Water Resources Review Committee of the Colorado Legislature." Alamosa, CO: Rio Grande Water Conservation District, Sept. 26, 2013.

Vandiver, Steve, General Manager, Rio Grande Water Conservation District. Interview by D. Stiller, May 12, 2016.

———. Interview by D. Stiller, March 11, 2014.

———. Interview by D. Stiller, March 15, 2013.

Vélez de Escalante, Silvestre. "Diary and Itinerary: The Story of the Escalante Expedition to the Interior Basin, 1776." *American Journeys*. Ed. Herbert Eugene Bolton. Utah State Historical Society, 1950. Accessed July 9, 2012. Online facsimile edition at http://www.americanjourneys.org/aj-106/print/index.asp.

Walters, Bruce C. "A Brief History of the Public Trust Doctrine in Colorado: Arguments Made for and Against its Application." *University of Denver Water Law Review* (June 10, 2015). http://duwaterlawreview.com/2015/06/.

Washington State Department of Ecology. *Washington State Water Law: A Primer*. Publication #WR 98-152 (Revised July 2006), Lacey, WA: Washington Department of Ecology, 2006.

Waskom, Reagan. "A Tale of Two Rivers: Groundwater Management in the South Platte and Rio Grande Basins." *State of the Basin Symposium at Adams State University*. Alamosa, CO: Salazar Rio Grande del Norte Center and Department of Biology and Earth Sciences, 2019.

Waterscape International Group. "Fact Sheet #005-01: California Water Rights and the Public Trust Doctrine." *Waterscape*, Jan. 18, 2018. waterscape.org/pubs/factsheet_waterright/FS-CaliforniaWaterRights.htm.

Webb, Walter Prescott. *The Great Plains*. Boston: Ginn and Company, 1931.

Wells, A. J. *Government Irrigation and the Settler: California, Oregon, Nevada and Arizona, Including a Description of the Imperial Valley Project*. San Francisco: Passenger Department, Southern Pacific, 1910.

West, Elliott. *The Contested Plains: Indians, Goldseekers, and the Rush to Colorado*. Lawrence, KS: University Press of Kansas, 1998.

Wheeler, George M., 1st Lieutenant. "Economical Features of Central Colorado, [Atlas Sheet No. 61(B)]. Issued March 3, 1876. Weyss, Herman & Lang, Del. Expeditions of 1873, 1874 & 1875 Under the Command of 1st Lt. George M. Wheeler, Geographical Surveys West of the 100th Meridian," 1876. *David Rumsey Historical Map Collection*. March 3. Accessed Apr. 5, 2014. http://www.davidrumsey.com/luna/servlet/detail/RUMSEY~8~1~346~30053.

———. "Land Classification Map of Part of Southwestern Colorado, Atlas Sheet 64(D)." Corps of Engineers, U.S. Army. *U.S. Geographical Surveys West of the 100th Meridian*, 1877. Washington, D.C. Accessed Apr. 10, 2013.

Wheeler, George M., 1st Lieutenant, Corps of Engineers, U.S. Army. "Economic Features of Parts of South'n Colorado & North'n New Mexico, [Atlas Sheet No. 69(B)]. Weyss, Lang & Herman Del. Expeditions of 1874–75 & 1877, Under the Command of 1st Lt. George M. Wheeler, Geographical Surveys West of the 100th Meridian," 1878. *David Rumsey Historical Map Collection*, Jan. 8. Accessed Apr. 5, 2014. http://www.davidrumsey.com/luna/servlet/detail/RUM SEY~8~1~367~30074.

———. "Economic Features of Parts of Southern Colorado and Northern New Mexico, [Atlas Sheet No. 70(A)]. Weyss, Herman & Mahlo Del. Expeditions of 1874, 1875 & 1876, Under the Command of 1st Lt. George M. Wheeler, Geographical Surveys West of the 100th Meridian," 1877. *David Rumsey Historical Map Collection*, May 7. Accessed Apr. 5, 2014. http://www.davidrumsey.com /luna/servlet/detail/RUMSEY~8~1~372~30079.

———. "Land Classification Map of Part of South Western Colorado, [Atlas Sheet No. 61(D)]. Weyss, Herman & Lang, Del. Expeditions of 1873, 1874, 1875 & 1876 Under the Command of 1st Lt. George M. Wheeler, Geographical Surveys West of the 100th Meridian," 1878. *David Rumsey Map Collection*, Jan. 10. Accessed Apr. 6, 2014. http://www.davidrumsey.com/luna/servlet/detail/RUM SEY~8~1~353~30060.

Worster, Donald. *Rivers of Empire: Water, Aridity, and the Growth of the American West*. New York: Oxford University Press, 1985.

Wulf, Andrea. *The Invention of Nature: Alexander von Humboldt's New World*. New York: Vintage Books, 2015.

Index

Page numbers for definitions are in boldface.

About the Author

DAVID STILLER is a former hydrologist, environmental consultant, university teacher, and nursery and farm owner and operator. He owned and managed a consulting firm specializing in mining hydrology and hazardous waste remediation, taught at Colorado Mesa University and the University of Colorado at Boulder, and served on a company's board of directors, attending to the water needs of irrigators of more than ten thousand acres of farmland in western Colorado. His earlier writings include *Wounding the West: Montana, Mining, and the Environment*. He holds a BS and MA from the University of Wyoming and a PhD from the University of Calgary. He lives near Durango, Colorado.